D0024910

More Chemistry and Crime

From Marsh Arsenic Test to DNA Profile

Samuel M. Gerber, *Editor*
Richard Saferstein, *Editor*

American Chemical Society
Washington, DC

Library of Congress Cataloging-in-Publication Data

More chemistry and crime : from marsh arsenic test to DNA profile / Samuel M. Gerber, editor, Richard Saferstein, editor

p. cm.

Includes bibliographical references and index.

ISBN 0-8412-3406-X

1. Chemistry, Forensic. 2. Criminal investigation. 3. Crime laboratories. I. Gerber, Samuel M., 1920– II. Saferstein, Richard, 1941–

HV8073.M763 1997
363.25′6—dc21 97-29913
CIP

The paper used in this publication meets the minimum requirements of American National Standard for Information Sciences—Permanence of Paper for Printed Library Materials, ANSI Z39.48-1984.

PRINTED IN THE UNITED STATES OF AMERICA

About the Editors

SAMUEL M. GERBER is a specialist in the chemistry and technology of dyes and their intermediates. He received a B.S. degree in chemistry from the City College of New York and an M.S. and Ph.D. from Columbia University. Most of his professional career was with the American Cyanamid Company, where his positions included Chief Chemist of Dyes and Intermediate Manufacturing and Manager of Dyes and Chemicals R&D. At present, he is a consultant in the field of dyes and related products and a principal in Color Consultants. His professional publications and patents are largely in the field of dyes and intermediates; they include an extensive compilation of Soviet contributions on azo and diazo chemistry. He is editor of *Chemistry and Crime* (American Chemical Society, Washington, DC, 1983).

Gerber has held offices both at the national and local levels of the American Chemical Society and is past president of the Royal Society of Chemistry (U.S. Branch). Other societies include the American Association of Textile Chemists and Colorists and the Society of Dyers and Colorists (England).

Gerber's interest in forensic chemistry originated with Sherlock Holmes. He is a member of the Baker Street Irregulars and related groups.

He has widely presented his lecture "Sherlock Holmes, Chemist". He is married to Barbara Brand Gerber, a media specialist, and they share three children and two grandchildren.

RICHARD SAFERSTEIN is a forensic science consultant. He retired as Chief Forensic Scientist of the New Jersey State Police Laboratory in 1991 after serving for 21 years. He received B.S. and M.A. degrees from the City College of New York. He received his Ph.D. degree in chemistry from the City University of New York. Prior to his coming to the New Jersey State Police in 1970, he was employed as a forensic chemist with the Treasury Department (1964–1968) and served as an analytical chemist with Shell Chemical Company (1969–1970).

Saferstein is the author of more than 30 technical papers covering a variety of forensic topics. He also has written a widely used textbook on the subject titled *Criminalistics: An Introduction to Forensic Science, 6th edition* (Prentice Hall, Englewood Cliffs, NJ, 1998) and has edited *Forensic Science Handbook,* Volumes I–III (Prentice Hall, Englewood Cliffs, NJ, 1982, 1988, 1993), popular reference texts dealing with important forensic science topics. He is a member of the American Chemical Society, American Academy of Forensic Science, International Association for Identification, Forensic Science Society of England, Canadian Society of Forensic Scientists, Northeastern Association of Forensic Scientists, Northwestern Association of Forensic Scientists, and Mid Atlantic Association of Forensic Scientists.

Contributors

DAVID H. BING received a Ph.D. in microbiology from Case Western Reserve University. After postgraduate training in immunochemisty at the University of California at Berkeley, he moved to Michigan State University, where he became tenured as an associate professor in the Department of Microbiology and Public Health. He moved to the Center for Blood Research, a Harvard Medical School affiliated research institute, in 1973. In 1985 he became Director of Clinical Testing at CBR Laboratories, which has been performing DNA-based human identity testing for disputed parentage and forensic analyses since 1989. Dr. Bing is a representative of the Human Identity Trade Association to the Technical Working Group on DNA Analysis Methods. He teaches forensic sciences at Northeastern University and human genetics at Brandeis University.

PAUL M. DOUGHERTY received a B.S. in criminalistics in 1956 from the University of California and a Juris Doctor in 1971 from Lincoln University. For the past 37 years, in a career which has utilized both law and criminalistics, he has had unusual experiences as an expert witness. An enthusiastic antique microscope and arms collector, he spends as much time as possible with others who enjoy the same interests.

ROBERT H. GOLDSMITH has a Ph.D. in chemical pharmacology from the University of Maryland and is Professor of Chemistry at St. Mary's College of Maryland. His interests are nutrition (especially trace elements) and the history of chemistry. He is a past chair of the ACS Division of the History of Chemistry and a fellow of the American College of Nutrition, as well as a certified nutrition specialist. Sherlock Holmes and Lord Peter Wimsey are two mystery favorites of his.

CHARLES R. MIDKIFF, JR., received an M.S. degree in chemistry in 1971 and an M.S. in forensic science in 1974, both from George Washington University. He has been a forensic chemist with the Bureau of Alcohol, Tobacco, and Firearms laboratory since 1971 and an adjunct professor in the Department of Justice, Law, and Society at American University since 1977.

NICHOLAS PETRACO received an M.S. in forensic science from John Jay College of Criminal Justice in 1979. During a 22-year career as an officer for the New York City Police Deptartment, he was assigned to the Crime Laboratory as a Detective/Criminalist. Mr. Petraco was involved in thousands of criminal investigations during his tenure at this laboratory, and he has testified in numerous high-profile cases such as Murder at the Metropolitan Opera, The CBS Homicides, The Central Park Jogger, and The Preppy Homicide, to name just a few. An avid mystery fan, he enjoys both reading and writing detective stories.

ROBERT SHALER is the Director of the Department of Forensic Biology at the Office of Chief Medical Examiner in New York City. After receiving his Ph.D. in biochemistry in 1968 from the Pennsylvania State University, he found himself attracted to forensic sciences, which he practiced in Pittsburgh and later in New York. When not analyzing crime scenes and worrying about DNA testing in homicides and sexual assaults, he is writing and fixing antique clocks. He is also an avid glass collector and antiques dealer.

JAY SIEGEL received a Ph.D. in chemistry from George Washington University in 1976. He has worked as a forensic chemist with the Virginia Bureau of Forensic Sciences and as a professor of chemistry and forensic science at Metropolitan State College in Denver. He has been on the faculty of the School of Criminal Justice at Michigan State University for the past 17 years, where he currently holds the positions of Professor of Forensic Science and Assistant Director. Dr. Siegel has testified as an expert witness in forensic evidence more than 200 times in 12 states and in federal and military courts.

MARINA STAJIC received a B.S. in chemistry from the University of Novi Sad in Yugoslavia and a Ph.D. in forensic toxicology from the University of Maryland, Baltimore. She worked in toxicology for the Commonwealth of Virginia for nine years. Since 1986 she has been the Director of the Toxicology Laboratory in the Office of the Chief Medical Examiner in New York City. She is a member of The Adventuress' of Sherlock Holmes and of the Baker Street Irregulars.

DAVID A. STONEY received a B.S. and M.Ph. in chemistry and criminalistics and a Ph.D. in forensic science from the University of California at Berkeley. He served as the Director of the Forensic Science Department at the University of Illinois at Chicago from 1985 to 1992. He is currently the Director of the McCrome Research Institute in Chicago.

Contents

Preface

Since the publication of *Chemistry and Crime* in 1983, a number of significant advances have occurred in the practice of forensic science to warrant the publication of a second volume in this series. A spate of criminal acts and high-profile trials also have served to exalt the role that forensic science plays in criminal investigation. *More Chemistry and Crime* explains current and emerging forensic science technologies, yet also examines their underpinnings by reviewing the historical development of forensic science philosophies and practices. A wide spectrum of techniques are addressed in this fashion, ranging from the microscope to DNA.

The editors are particularly fortunate to have within the text a chapter authored by the late Ralph Turner, one of the recognized founders of criminalistics. His death in 1994 terminated an illustrious career in forensic science spanning five decades.

Forensic science subjects cut a wide swath, and the topics contained within this book reflect their number and diversity. Included are chapters dealing with arson analysis, forensic serology, DNA analysis, and forensic toxicology. A thorough exploration of past practices of forensic microscopy illustrates how history can serve as a guidepost for current and future applications of this discipline. Siegel's historical treatment of the role of the expert witness illustrates the unique position that the scientist plays in offering testimony in the courtroom. Whether the current adversarial system is an appropriate forum for eliciting complex technical information from scientists for evaluation by lay jurors remains an area of intense debate.

Finally, this volume delves into the ever-greater role that forensic science plays in fictional portrayals of criminal investigations. The public has an insatiable appetite for seeing, hearing, and reading about the role of science and technology in crime solving. We hope that, after completing this book, the reader will conclude that fact is not far removed from fiction.

SAMUEL M. GERBER
RICHARD SAFERSTEIN

Dedication

To Ralph F. Turner
1917-1994

Outstanding pioneer forensic scientist and educator,
architectural buff, enthusiastic Sherlockian and
devoted husband and father.
A true Renaissance man... our good friend.

RALPH F. TURNER was born in Wisconsin in 1917 and educated at the University of Wisconsin, University of Southern California, Harvard, and Yale. His pioneering efforts in forensic science led to the establishment of the first police laboratory in Kansas City, Missouri and the founding of the American Academy of Forensic Science. After serving as laboratory supervisor of the Kansas City Police Department (1939–1947), he joined Michigan State University, where he established the criminalistics program and was a professor of criminalistics (1947–1981). His forensic science achievements, particularly in ballistics and alcohol detection and measurement, soon gave him an internal reputation. He taught, developed forensic education, and helped established forensic laboratories in South Vietnam, Taiwan, Guam, Saudi Arabia, and England.

As an educator and scientist he established standards of rigorous proof that gave firm credibility to the courtroom testimony of the forensic expert witness. We are all in his debt.

He was a member of several professional societies and received many awards, including Michigan State University Distinguished Faculty Award and the Bruce Smith Award of the Academy of Criminal Justice Sciences.

His interests were wide. He was a Sherlock Holmes aficionado; he took a great interest in architecture, particularly that of Frank Lloyd Wright; and he was an art lover who took up oil painting in his later years.

Ralph Turner died in 1994, leaving his wife of 53 years, Arnella, three children, and four granddaughters.

1

Historical Perspectives in Forensic Science

Ralph F. Turner

Perspective is understood to mean the human capacity to view things in their true relationships or relative importance. But, as Henry Adams observed, these relationships and their importance are altered by time and experience.

Readers of this book will find numerous references to important historical events and individuals in the chapters dealing with various aspects of forensic science. Any attempt to describe all of the developments in this field would require more space than is appropriate here, but devoting a single chapter to historical perspectives suggests that the writer may take liberties with the assignment—selective liberties that reflect my personal perspective.

My overview spans roughly 50 years, 1939 marking the first time I testified in court (albeit in a minor firearms identification case) and 1989 being the year I ceased accepting requests for forensic consultation. Putting forensic science into some kind of perspective during that span permits me, I think, the opportunity to describe the developing state of the art and to reflect on the changing role of forensic science in the administration of justice.

Making Clues Tell the Story

The development of forensic science has always been closely related to the needs of the criminal investigator. This is best exemplified in the work of Hans Gross. Between 1883 and 1914, six editions of his *Handbuch fur Untersuchungsrichter als System der Kriminalistik* appeared; the first English adaptation was published in 1906 in Madras, in a translation *(Criminal Investigation)* by John Adam and J. Collyer Adam. Subsequent editions followed, with a fifth version appearing in 1964. In my view, this work by Gross is the basic and fundamental guideline for criminal investigation and the subsequent use of forensic science. I included the following quote from the 1934 edition of *Criminal Investigation*, edited by Norman Kendal, in my book *Forensic Science and Laboratory Technics:*[1]

> The trace of a crime discovered and turned to good account, a correct sketch, be it ever so simple, a microscopic slide, a tattooing, a restored piece of burnt paper, a careful survey, a thousand more material things are all examples of incorruptible, disinterested, and enduring testimony from which mistaken, inaccurate, and biased perceptions, as well as evil intention, perjury, and unlawful co-operation, are excluded. As the science of Criminal Investigation proceeds, oral testimony falls behind and the importance of realistic proof advances: "circumstances cannot lie"; witnesses can and do.

My evaluation of the importance of Gross's contributions led me to distill my practical experience into lectures on criminal investigation that included two hypothetical statements: (1) It is impossible to commit a crime without leaving a clue. (2) If *all* evidence is collected and interpreted *correctly* it will lead to the perpetrator. This concept has been described in part in a master's thesis by Wilbur Rykert.[2] I feel that an interpretation of Gross is as valid today as it was a hundred years ago. Gross included in his discussion of criminal investigation the value of enlisting the aid of scientists who could interpret wounds, telltale markings of a struggle, the presence and effect of poisonous substances, inferences gained from studying the patterns of drops of blood on a floor, the value of understanding modus operandi, and the importance of applying all scientific and technical information to the solution of a criminal investigation.

Concurrent with the work of Gross in Austria, Arthur Conan Doyle presented a fictional adaptation of the scientific method in criminal investigation through the medium of the enduringly popular Sherlock Holmes stories. Insofar as I have been able to research the matter, these two gentlemen, Gross the examining magistrate and Doyle the trained physician, never met. However, their descriptions of methods of collecting and interpreting forensic evidence were strikingly similar.

The period shortly before and after the turn of the twentieth century included important work by a number of other Europeans, including Edmond Locard, Alphonse Bertillon, and Sir Edward Henry, all of whom made contributions in the area of personal identification. Henry's development of a system of classification of fingerprints is, in fact, still in use today. The incorporation of the scientific method in criminal investigation and in subsequent courtroom trials became evident in the United States at this time. Forensic pathology and toxicology were introduced in 1877 in the first official medical examiner's office, in the state of Massachusetts, and in a similar office in New York City in 1915. Firearms identification attracted attention during the trial of Charles Stielow in New York State in 1915. The first decade of the twentieth century saw the use of document examiners in many investigations, both civil and criminal. As new information was developed in the field of personal identification, fingerprint pattern classification became the basic means for establishing identity. The importance of blood grouping was also recognized as a valuable tool in linking a suspect with a particular crime scene and also in resolving cases of disputed paternity. Methods of interrogation were enhanced with the development of the polygraph and, to a limited extent, the use of "truth serum".

With the spin-off of technology developed during and after World War II, forensic science acquired a cohesiveness not apparent prior to this time. The use of scientific expertise was practiced almost exclusively by individual consultants, except for those employed in a few police identification bureaus or scientific laboratories. The Scientific Crime Detection Laboratory at Northwestern University was founded after the St. Valentine's Day massacre in Chicago in 1929, and the FBI laboratory was established a few years later. For greater detail on the history of forensic science in the United States, see Duayne Dillon's *A History of Criminalistics in the United States, 1850–1950.*[3]

Early Twentieth Century Forensic Science

If one reflects on the state of forensic science in its early days in the United States, several distinguishing characteristics come to mind. First, with the exception of the physician trained in pathology and the analytical chemist specializing in toxicology, the typical forensic scientist was a unique individual, frequently self-taught in a specialty, who applied scientific techniques to aid in the solution of a criminal investigation. There were no boards of review or examinations to ensure the professional competence of the expert, nor were there organizations to provide a forum for the exchange of pertinent information. The *American Journal of Police Science*, published by Northwestern University Law School, was initiated in 1930 and incorporated with the *Journal of Criminal Law, Criminology, and Police Science* two years later.

Second, when forensic experts other than medical and toxicological witnesses testified in court, there seldom was rebuttal testimony by authorities of equal stature. The celebrated trial in 1935 of Bruno Hauptman, charged with the kidnapping and murder of Charles Lindbergh, Jr., brought no sophisticated challenges to the testimony of Albert Osborn, document examiner, or of Arthur Koehler, wood identification expert from the U.S. Forest Products Laboratory on the campus of the University of Wisconsin. It was my privilege while an undergraduate student and protege of J. H. Mathews, chairman of the chemistry department at the university, to hear Koehler and Mathews discuss elements of the Lindbergh case after the trial. As the number of forensic scientists increased, beginning in the 1950s, it was inevitable that expert testimony would no longer go unchallenged. It appears at present that California probably has the largest number of forensic scientists testifying in court or reviewing evidence for litigants on either side of criminal or civil cases, a view undoubtedly colored by my experience in the Midwest during the period 1939–1954 and subsequent opportunities to examine conditions in California in 1955 and later.

Third, with few exceptions, there was little opportunity for formal training in the first half of the twentieth century in any of the fields associated with forensic science. Aside from the aforementioned medical examiner offices in Massachusetts and New York, forensic work done by physicians or chemists was usually on an ad hoc or extracurricular basis. The first Department of Legal Medicine was formed at Harvard University shortly after World War II, largely due to the generous financial support of Mrs. Francis G. Lee, its wealthy benefactor, and to the pioneering work of Alan Moritz, its first director. Curricula leading to graduate degrees in forensic toxicology were disorganized at best and students had to search diligently to find suitable programs if they wished to pursue a career in this field. Formal training in police administration began around 1935 with the programs at San Jose and Berkeley in California and at Michigan State University. Paul

Kirk introduced the program in criminalistics at the University of California and R. Turner developed a similar program at Michigan State in 1948.

Fourth, probably one of the outstanding features of the use of forensic scientists in the courtroom was the manner in which the expert witness presented testimony and interpreted the evidence to the judge or jury. The British physician Sir Bernard Spillsbury employed dramatic simulation and reenactment to demonstrate strangulation being masked by drowning in a bathtub. Expert witnesses presented illustrated lectures or demonstrations, sometimes highly technical, to the jury in the effort to aid the jurors to concur with the witness's testimony.

As his understudy, I once accompanied J. H. Mathews to a trial, both of us loaded down with a portable bullet recovery box, a comparison microscope, a 3.25-by-4-inch lantern slide projector, a screen, a stereopticon viewer along with stereo photomicrographs, plus conventional 8-by-10-inch prints for each juror to view. Testimony by Mathews could take one to two days and usually included a lecture on how microscopic imperfections in a rifled gun barrel are transferred to the surface of the bullet as it passes through the barrel. Test shots would be fired in the courtroom and jurors would be invited to peer through the comparison microscope and understand how the expert witness came to the conclusion that the fatal bullet was fired from the gun found in possession of the suspect at the time of arrest. Numerous photographs were always produced to support the testimony of the witness. Melvin Belli, the well-known California attorney, was an ardent supporter of the use of "demonstrative evidence". Such productions were the stuff that made headlines and, as noted earlier, were seldom challenged in a meaningful way.

Such testimony is in marked contrast to attitudes reflected today by some forensic witnesses, namely that too many photographs and too much detailed information provide opposing counsel with opportunities for confusing and distracting cross-examination—usually confusing for the jury and sometimes embarrassing for the witness. This attitude was brought to my attention in 1938 by Sergeant Harry Butts of the New York City Police Department Scientific Laboratory. He was very firm in his conviction that it was not necessary for him to bring photographs of his firearm identification work into the courtroom. His statement about the nature of his findings, he insisted, should be sufficient. Given my own tutelage under Mathews, I preferred to bolster expert opinion with as much graphic and illustrative evidence as possible. In retrospect, I feel that expert witnessing has evolved from a state of innocence manifested by the sincerity of pioneering experts to an era of cynicism wherein jurors and citizens alike are caught up in the technological maze of our confrontational system of justice. To be sure, miscarriages of justice occurred in those early days (such as the Charles Chaplin paternity suit), and injustices are suffered today by those defendants unable to enlist the services of skillful attorneys or expensive expert witnesses.

Visitors to forensic laboratories today may note a striking similarity in the appearance of the work space. Instrumentation dominates the scene, both in the United States and abroad. My photographs of laboratories that I have visited recently in Korea, Taiwan, England, and California could easily be interchanged; the instruments and the analytical procedures employed in all of them are similar. In contrast, forensic laboratories in the 1930s and earlier were very different from one another.

At that time, instruments were handcrafted for special adaptation to forensic problems. For example, the optical bridge, credited in the United States to Chamot and Mason, was adapted by Calvin Goddard to create the comparison microscope, used for bullet and cartridge examinations. Luke May in Seattle, Washington, developed an extremely large and unwieldy floor-model microscope-camera, which he dubbed the Magnascope; it also was used for comparative examinations of evidence. Goddard also designed the helixometer for examination of the interior of gun barrels. Mathews built a comparison camera that even today has a number of practical and unique features. Moreover, his rifling meter for measuring class characteristics of firearms is a one-of-a-kind tool. The stereo photomicrographs referred to earlier were made with a custom-built camera. John Larson and Leonarde Keeler did the pioneering work in developing the polygraph. John Davis constructed the Striagraph in the 1940s, although an earlier version employing the same principle was built by Mathews and Turner in 1938. Rolla Harger developed the Drunkometer at Indiana University Medical School; it was brought to national attention during the first extensive drinking-and-driving field study in Evanston, Illinois, by Northwestern University Traffic Institute in 1935.

These were heady days for workers who were introducing new techniques to be used in criminal investigation and courtroom testimony. Restoration of obliterated serial numbers on firearms claimed widespread attention when a new technique was introduced by technicians in the Northwestern University Scientific Crime Detection Laboratory that validated the identification of the weapon used in the daylight slaying of Jake Lingle, a well-known Chicago newspaper reporter. Truth serum (scopolamine or sodium amytal), adapted by the Texas obstetrician A. E. House for interrogation of prison inmates, also came into use by police investigators.

Developing Training and the Academy

World War II interrupted the orderly development of forensic science research, but beginning in 1945 a remarkable resurgence of activity in this field occurred on several fronts. Two events are significant from an historical standpoint. First, as stated earlier, there was little opportunity for formal training in forensic science prior to World War II. However, with scores of students turning to colleges and universities in the postwar era, a number of leaders in criminal justice administration recognized the need for college-trained police personnel. Programs in police administration suddenly were flooded, and in a few instances criminalistics programs were introduced at this time. Subsequently, employment of college-trained applicants by federal, state, county, and municipal law enforcement agencies became almost routine. The Department of Legal Medicine at Harvard University began its series of homicide investigation seminars, which set the tone for similar workshops around the country, many of which still flourish. The lawyer Melvin Belli and Hubert Winston Smith, professor at the University of Texas Law School, conducted many seminars to introduce participants to the use in trials of demonstrative evidence and forensic science information. Northwestern University Law School supplemented its widely recognized Prosecutors Short Course with a similar program for defense counsels. Here also recent advances in scientific criminal investigation were highlighted.

There was not, however, an abundance of literature to bolster these new advances. Gonzales, Vance, and Helpern's *Legal Medicine and Toxicology*,[4] Soderman and O'Connell's *Modern Criminal Investigation*,[5] and Hatcher's *Firearms Investigation, Identification and Evidence*[6] were some of the most widely used references. These American texts were supplemented by numerous British publications. In 1944, Charles C. Thomas of Springfield, Illinois, published LeMoyne Snyder's *Homicide Investigation*,[7] which marked the beginning of a new venture for publishers. With support from the federal government through its Law Enforcement Assistance Administration (LEAA), publications began to appear in impressive numbers, although not always of impressive quality. During the period 1945–1950, individual ingenuity and imagination were still at a high level, government bureaucracy had not yet overwhelmed the practitioners, and forensic science was becoming an integral part of the justice process rather than being a specialty reserved for a few headlinemaking cases.

The second significant event in the immediate postwar period was the formation of the American Academy of Forensic Sciences. With the growth of the number of forensic laboratories in the United States and the assembling of a body of knowledge unique to the field, it was only a matter of time before the need became apparent for a professional forum in which to disseminate and share that evolving knowledge. Thanks to the foresight, determination, and perseverance of R. B. H. Gradwohl of St. Louis, Missouri, the first American Medico-Legal Congress was held in that city January 19–21, 1948, at which 36 scientific papers were presented. Sessions were held at the training academy of the police department; Gradwohl personally underwrote most of the incidental expenses. There had been several attempts to develop a forensic society in the eastern part of the country, but this was the first such effort to succeed.

A committee appointed by Gradwohl, consisting of LeMoyne Snyder, Leonarde Keeler, Charles C. Thomas, Sidney Kaye, and Orville Richardson, met several times and submitted two reports to the general assembly. To establish a forensics society, they recommended a national scope and inclusion of "all divisions of science useful to the legal process to promote justice". They called for Gradwohl to appoint a committee "to sound out thought throughout the United States of all scientists, lawyers and jurists who would be of substantial assistance in the attainment of our purpose, and obtain from them suggestions and assistance in the formation of a national medicolegal society or national institute of law-science relationships". The committee then would call another general convocation to establish the society.

A steering committee of 24 individuals representing a cross section of the law-science community met at the Hotel Pierre in New York City on October 18, 1948, to implement those suggestions. An organizational meeting of the American Academy of Forensic Sciences was subsequently held at Northwestern University Law School in Chicago, January 26–28, 1950, with the third meeting of the academy being held at the Drake Hotel in Chicago, March 1–3, 1951. The 50th anniversary meeting is scheduled for Chicago in 1998. I had the privilege of serving as interim secretary 1948–1950, secretary-treasurer 1950–1955, and as chairman of the Police Science Section 1951–1952.

Two volumes of the *Proceedings*[8] were printed in 1951 and again in 1952, followed by publication of the *Journal of Forensic Sciences* with Samuel

Levinson as its first editor.[9] The *Journal*, now in its 37th volume, is published bimonthly, with over 6,000 copies going to readers around the world.

How Far We Have Come

It was my good fortune to be seated next to Alex Weiner at the evening banquet of the Medico-Legal Congress in St. Louis in 1948. He was the immunologist with the New York City Medical Examiners Office and the discoverer of the Rh factor in human blood. He was very optimistic about the future of personal identification by means of blood analysis and felt that in the not-too-distant future the science of blood identification would be as accurate as fingerprint examination in determining identity. One need merely compare the state of the art in the time of Hans Gross and Conan Doyle (human blood groups had not yet been discovered) with the remarkable prominence that DNA examinations enjoy today. Considerable progress has been made toward the realization of Weiner's prediction. It is also heartening to know that DNA testing is now being used to reexamine evidence in some sex crime cases and, where indicated, miscarriages of justice are being corrected.

Though improvements in the techniques of fingerprint identification per se have not been as dramatic as those in the field of forensic immunology, the application of computer technology has speeded up the search of massive fingerprint card collections in a most astounding manner. The Automated Fingerprint Identification System (AFIS) enables technicians to search thousands of fingerprint cards in a matter of minutes, in addition to being networked with other identification bureaus on a regional basis. Using a similar technique, the FBI has recently begun a trial run of its Drugfire program, in which surface features of fired cartidges are put into computer files and shared with firearms identification units on a regional basis.

Software programs for computerizing criminal investigation have been developed. At least 20 companies offer programs that can be used by investigators or task forces. These programs include records management, fingerprint classification, automatic searching, computer-aided sketching of suspects, and case management.

Improvements in analytical techniques, more of which are becoming nondestructive, have reached astonishing parameters with more NASA-generated technologies being transferred to the forensic laboratory. Caution must still be exercised, however. In a criminal investigation, it is the investigator who must always establish that an event (crime) has occurred, understand the modus operandi, discover a motive, connect the perpetrator to the event through some form of personal identification, and marshal proof to meet the demands of the law.

When scientific evidence is introduced as part of the trial procedure, it must be interpreted by the expert witness. At this point the *interpretation* becomes critical, for evidence falls into one of two categories: objective or subjective. Much of the forensic evidence introduced in today's trials falls somewhere in between (quantitative analysis of a drug is objective and disputed diagnoses of insanity may be subjective) and the burden of the interpretation of the expert testimony is then shifted to the jury.

A telling description of criminal investigation is the subtitle of James Osterburg's 1992 book *Criminal Investigation: A Method for Reconstructing the Past.*[10] In my opinion, all criminal investigations consist of procedures common to those used by archaeologists, who may deal with events hundreds of thousands of years old or, in the case of urban archaeology, be involved with artifacts of a more recent past. With the exception of well-known inquiries into historical cases, such as when arsenic was detected in samples of Napoleon's hair, criminal investigators and forensic scientists are generally involved in reconstructing events that occurred only minutes, hours, days, or weeks ago. However, the scientific methods used by archaeologists, criminal investigators, and forensic scientists are remarkably similar.

Time and experience have refined both the methods and the perspectives of the practitioners of these disciplines.

References

1. Gross, H. *Criminal Investigation;* Kendal, N., Ed.; Sweet and Maxwell: London, 1934. Turner, R. F. *Forensic Science and Laboratory Technics;* Charles C. Thomas: Springfield, IL, 1949.
2. Rykert, W. L. M.S. Thesis, Michigan State University, 1985.
3. Dillon, D. J. *A History of Criminalistics in the United States;* University Microfilms International: Ann Arbor, MI, 1977.
4. Gonzales, T. A.; Vance, M.; Helpern, M. *Legal Medicine and Toxicology;* D. Appleton Century: New York, 1940.
5. Soderman, H.; O'Connell, J. J. *Modern Criminal Investigation;* Funk and Wagnalls: New York, 1935.
6. Hatcher, J. S. *Firearms Investigation, Identification and Evidence;* Small Arms Technical: Marines, Onslow County, NC, 1935.
7. Snyder, L. *Homicide Investigation;* Charles C. Thomas: Springfield, IL, 1944.
8. *Proceedings of the American Academy of Forensic Sciences;* Dutra, F. R., Turner, R. F., Eds.; Edwards Bros.: Ann Arbor, MI, 1952; Vol. 1.

9. *Journal of Forensic Sciences;* American Society for Testing and Materials (ASTM), 1916 Race Street, Philadelphia, PA; 19103.
10. Osterburg, J. W.; Ward, R. H. *Criminal Investigation: A Method for Reconstructing the Past;* Anderson: Cincinnati, OH, 1992.

Suggested Readings

Doyle, A. Conan. *Stories of Sherlock Holmes*; Harper and Brothers: New York, 1904.

Hudzid, John K. *Federal Aid to Criminal Justice*; National Criminal Justice Association: Washington, D.C., 1984.

Kirk, Paul L. *Crime Investigation*; Interscience Publishers: New York, 1953.

O'Hara, Charles E.; Osterburg, James W. *An Introduction to Criminalistics*; Macmillan: New York, 1949.

Palenik, Skip. *Microscopic Trace Evidence: The Overlooked Clue*. Part II, *Max Feri: Sherlock Holmes with a Microscope*; McCrone Research Institute: Chicago, 1982.

Peterson, J. L. et al. "The Capabilities, Uses and Effects of the Nation's Criminalistics Laboratories" *Journal of Forensic Sciences*, **1985,** *30(1)*, pp 10–23.

Peterson, J. L., "Ethical Issues in the Collection, Examination and Use of Physical Evidence", *Forensic Science*, 2nd Ed. American Chemical Society: Washington, D.C., 1986.

Stead, Philip J. *Pioneers in Policing*; Patterson Smith: Montclair, NJ, 1977.

2

Forensic Biology

A Walk Through History

Robert C. Shaler

Science advances steadily, sometimes even rapidly, but applying it to the real world can take years. As the breadth of scientific knowledge increases, so too should the specificity of scientific testimony. The legal system is challenged to deal with scientific evidence. Errors in judgment no doubt have occurred as lawyers and judges, nonscientists to be sure, render decisions regarding the admissibility of scientific evidence. Sometimes bad science is admitted, such as the use of neutron activation analysis to determine metal levels in hairs as a basis of individual identification. In almost all circumstances, however, these errors are corrected as the adversarial system applies its rules of evidence regarding the admittance of novel scientific techniques. In fact, the courts do handle new scientific issues with relative success.

Forensic biology has a long history, and, on reflection, its influence on the world's criminal justice system is apparent. The types of conflicts visible today as scientists wrestle with the admissibility and reliability of the latest frontier in forensic biology, DNA profiling, are no less evident than they were in earlier years.

Disciplines of Forensic Biology

The forensic biologists of earlier times were often chemists, actually toxicologists, and some were cytologists or even physicians. As immunology became the hot topic, the expertise of forensic biologists expanded and they began using immunological tools, becoming applied immunologists. This became the status quo for more than 60 years until modern biochemical methods became available and forensic biologists added biochemistry to their repertoire of tricks, becoming applied biochemists. Now, more than 20 years after the introduction of biochemical separation methods, they are becoming applied molecular biologists because it has become necessary to incorporate recombinant deoxyribonucleic acid (DNA) technology into their procedural arsenal. What is most interesting is that old tests have not all been abandoned. Many are still used and remain a part of the battery of tests used in modern forensic biology. Figure 1 shows approximately when different disciplines became integrated into the testing protocols of forensic biologists.

If we highlight the accomplishments of forensic biology by asking, for each major advance, what we can conclusively identify, then we would have a series of questions that would sequentially expand to include our current "hot scientific topic". The questions would evolve as: Is a reddish substance on the trousers of the suspect truly blood? Is it truly human blood? Is it human blood of a type matching the deceased? Is it human blood with a genetic profile of 15 antigenic and biochemical markers that match the deceased? And, most recently, is it human blood with a DNA profile identifying the deceased?

These questions suggest an historical categorization of forensic biology into arbitrary and overlapping periods of varying length based on scientific advances. These periods may be categorized as Classification, Identification, Differentiation, and Individualization; see Figure 2. *Classification* and

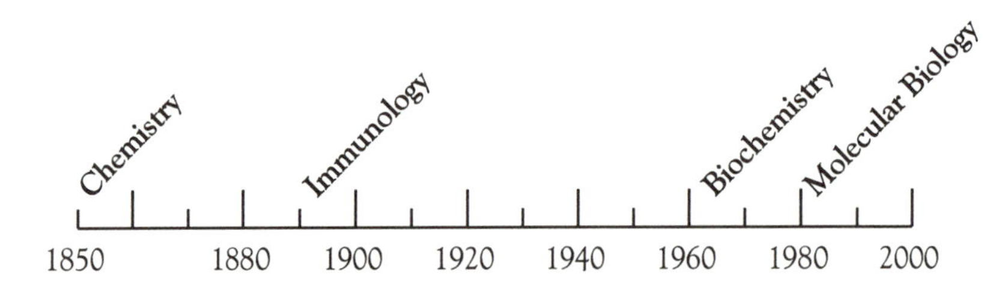

Figure 1. The Scientific Disciplines of Forensic Biology

Identification refer to the ability to categorize an unknown stain as blood and as human blood, respectively. *Differentiation* is the ability to tell the difference between blood samples of unrelated sources; *Individualization* identifies whose blood it is. Individualization has had two phases, the first being genetic profiling before DNA testing and the second being such profiling after DNA testing.

This chapter first examines the evolution of expertise required of forensic biologists and then describes the tools and methods used during each period.

Personnel in Forensic Biology Laboratories

The qualifications of forensic biologists are a question of contemporary importance, especially regarding legal challenges to the admissibility of scientific evidence. It is clear that current standards have evolved as science advanced.

The early tests were performed by chemists, physicians, or toxicologists. Catalytic tests are strictly chemical tests and, though a knowledge of chemistry is needed to understand the mechanism of the test, it is fair to say that the early chemists used the tests without a detailed knowledge of how they worked. That they worked properly under the conditions in which they were being used was all that was needed to provide useful information for police and legal investigators. The same is true for the immunological tests used to identify species or blood groups.

The biochemical tests brought in a new set of techniques between the late 1960s and the 1980s, coinciding with a period of rapid expansion for crime laboratories. Associated with the expansion was an influx of a new breed of scientist trained in biochemical methods. The mixture of expertise in the laboratories was becoming complex. Some of the laboratory person-

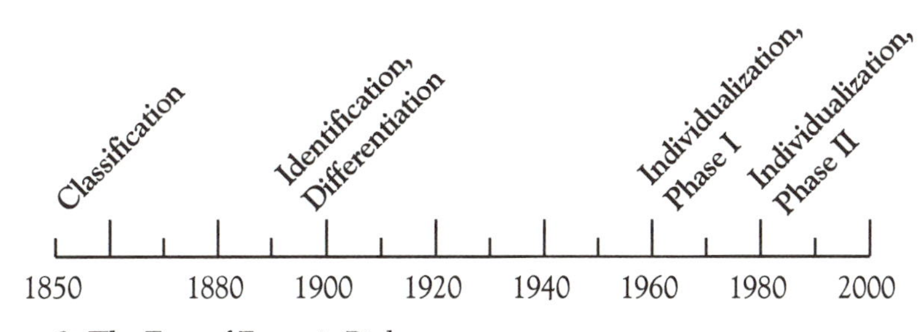

Figure 2. The Eras of Forensic Biology

nel from the early days of the 1950s and 1960s were familiar with blood identification, species determination, and blood grouping, but not with the biochemical techniques. Some laboratory systems, such as that of the New Jersey State Police, recognized this trend and created biochemistry units. These units continued to perform the old serological tests but also incorporated the new biochemical ones into their testing procedures.

The application of molecular biological techniques to forensic investigations has created similar changes. Forensic molecular biology or DNA sections are being created in laboratories. Some labs are adding in molecular biologists to their staffs. Other labs are training existing personnel in the new techniques.

Laboratory directors have realized that the field encompasses a broad range of disciplines and that to call these scientists serologists or forensic serologists is no longer appropriate. The forensic scientist who specializes in the analysis of biological evidence is now considered a forensic biologist, a term that better reflects the diverse testing required. The background training may be chemistry, forensic science, biochemistry, immunology, or molecular biology. Some existing staff members must educate "up"—that is, learn the new techniques and applicable theory; others must educate "down" by learning the old techniques and applicable theory.

The history of the forensic biology laboratory at the Office of the Chief Medical Examiner in New York City is an example of how laboratories have changed. At its inception in the early 1930s it was known as the Serology Laboratory; it was directed by Alexander Weiner, a physician. It retained this designation until 1990 when it was changed to the Department of Forensic Biology, housing two laboratories: the Forensic Biochemistry and Hematology Laboratory and the Forensic Molecular Biology Laboratory. The director is now a Ph.D.-level biochemist.

The title of the laboratory director has changed as well. Most forensic biology laboratories are public and thus are governed by civil service systems; consequently these titles tend to change slowly. In 1932, Weiner's civil service title was Bacteriologist. In 1978, the laboratory director's civil service title was changed to Research Scientist III with an in-house title of Chief Serologist. In 1990, the laboratory director was removed from civil service and given the managerial title, Director of Forensic Biology.

Ultimately, forensic biologists are concerned with the unequivocal identification of the sources of unknown biological specimens collected as a part of an investigation. Modern forensic biologists perform a series of tests that tend to mimic the history of the forensic biological sciences. These tests progress from chemical, immunological, and biochemical to molecular biological in an effort to attain the ultimate goal. This progression represents the current concept of the proper employment of test variations that have evolved as forensic biology has developed.

The progression of tests has been aided by an adversarial legal system that cannot be ignored because it has helped mold forensic science. The legal challenges to the admissibility of scientific evidence act as a buffer between good and bad science; the courts may seem slow to act, but the legal decisions rendered over time tend to correctly choose science that is most reliable.

With this in mind, the remainder of this chapter discusses the application of science as it has been used by forensic biologists in modern times. The specific tests described are those that have been used in the majority of United States forensic biology laboratories; specialized or rarely employed tests are not described except in instances where history should be preserved.

Era of Classification: Identification of Blood

Identifying an unknown substance on a suspect's clothing as human blood may have important inculpatory implications in criminal cases. The actual probative value of the evidence depends on the circumstances of each case. As the following case description illustrates, the presence of blood alone may be important. A pair of scissors was recovered from beneath an automobile. A witness, who later disappeared, pointed out the scissors to police officers who were dispatched to the scene. The victim was stabbed 10 times through three layers of clothing. Most of the wounds were superficial and only two of them broke the surface, penetrating to a depth of only 1/8 to 1/4 of an inch. No visible staining of the scissors was apparent, and presumptive testing gave weak results. The results were expected, but since the test was being used at the limits of its sensitivity, a conclusive interpretation regarding the presence of blood could not be made. A battle of experts ensued regarding the interpretation of the results based on the laboratory's report that blood was not present.[1]

Presumptive Tests: Chemical or Catalytic Tests

From an investigative standpoint, the unequivocal identification of an unknown dried substance as blood is an important first step in identifying the origin of biological evidence. At one time it was the first, final, and only step in the analytical scheme available for forensic biologists.

In fact, this relatively simple test has a stormy scientific history. Even 130 years after its first use, endless variations of the test are still fodder for cross-examination and can create a battle between expert witnesses.[1] As in most instances, the individual case details or lack thereof create situations for adversarial debate. Often, adversarial debates occur when important case issues rely on the scientific testing, and the opposition has the financial resources with which to challenge the scientific results or their interpretations.

The basis of all these tests for the presence of blood is the same: the peroxide-mediated oxidation of a variety of organic compounds, catalyzed by the iron in hemoglobin. The tests vary with the organic compounds used, the mode of incorporating peroxide in the reaction, and whether the test is employed in one, two, or three stages. The organic compounds are extensively conjugated ring systems that, upon oxidation, produce a colored or chemiluminescent product. Each variation has its proponents and opponents.

The first of these tests designated for medicolegal use was the Van Deen's, Day's, or "Antozone" test.[2] It exploited the ability of blood, in the presence of peroxide, to turn guaiacum blue. Developed in 1862, it became the most prominent of the original catalytic tests. Perhaps the first reported case using Van Deen's catalytic test involved the finding of positive results on a pair of washed trousers taken from an accused murderer.

As with most new scientific discoveries, the interpretation of the results of catalytic tests created controversy. Each interpretation had proponents and opponents. In London, A. S. Taylor regarded the test as "useful" in conjunction with other tests. He believed that negative results constituted reliable proof for the absence of blood but that positive results only gave reasonable certainty as to the presence of blood.[3] This view represented the opinion of the majority of scientists at the time. Whether or not the catalytic tests were conclusive as to the presence of blood caused debate. Some believed that, if the test had been employed properly, the stain was "surely blood"[4] or there was a "high probability" of blood being present.[5] Most scientists of the day believed that the test had value as a screening technique but did not put too much weight on a positive result as being conclusively blood;[6–8] others, however, believed that the only value of the tests was to ensure that stains were not blood.[9–11] A minority of scientists believed that the test had no value at all.[12–13]

Even the proper procedures for conducting the tests were not universally agreed upon, and as recently as 1976 forensic scientists were concerned with which of the most recent modifications should be used and how the test was to be employed.[14] It was known that false positive results could be obtained from a variety of oxidants.[15] These could be detected by employing a two-stage test where the organic compounds were added first to the substance being tested, followed by peroxide; a reaction without the addition of peroxide constituted a false reaction.

One of the most widely used of the catalytic tests employed the carcinogenic compound benzidine, *p*-diaminodiphenyl.[16] Its use has virtually been discontinued, being replaced with *N,N,N′,N′*-tetramethylbenzidine[17] or phenolphthalin.[16,18–20] Since the 1970s, the phenolphthalin (Kastle–Meyer) test has become the workhorse of catalytic chemical tests used for the presumptive identification of blood.

Another category of catalytic tests uses organic compounds whose oxidation product has chemiluminescent properties. The most commonly used of these reagents is luminol, 3-aminophthalhydrazide.[21–22] This reagent is preferred by crime scene units for ascertaining the location of "invisible" blood at crime scenes. It is particularly helpful in identifying the location of blood that supposedly has been cleaned.[23] After application of the reagent, the room is darkened, and the areas with luminescence may contain traces of dried blood. In these situations, care must be taken to account for the surface being sprayed with the reagent; certain alloys containing copper also react with the reagent, giving a false positive result.

Confirmation of Blood: Identification of Hemoglobin

It is almost axiomatic that the confirmation of the presence of blood requires that hemoglobin be conclusively identified. Classically, several methods have served as hemoglobin identifiers. Among these, only the microcrystalline tests seem to have withstood the test of time.

Microcrystalline Tests. Among the many ways to confirm the presence of blood, the most commonly used are microcrystalline tests; in these, microcrystalline derivatives of hemoglobin—hematin, hemin, and hemochromogen—are observed microscopically. The development of the crystal tests spans a period of more than 80 years, from 1853 to 1935.[24–30] Though the methods have remained virtually intact, the Takayama test for the identification of pyridine-hemochromogen crystals is the most commonly employed in today's forensic biology laboratories.

Spectrophotometric and Electrophoretic Tests. Hemoglobin identification also can be accomplished using ultraviolet and visible spectrophotometry. First suggested by Hoppe in 1862 as a means of blood identification, the method has not gained widespread popularity in American forensic biology laboratories.[31] The characteristic mobility of hemoglobin in electrophoretic systems using cellulose acetate[32–33] or isoelectric focusing procedures constitutes a definitive method for hemoglobin identification. This approach is not used routinely in most American forensic biology laboratories.

Immunological Methods. The use of antiserum, which reacts specifically with human hemoglobin, is another approach for identifying hemoglobin and simultaneously identifies the species.[34] The lack of highly specific anti–human-hemoglobin sera has prevented this technique from gaining popularity. With the production of monoclonal antibodies that are prepared against human hemoglobin, the principle of the technique can be employed more successfully using enzyme-linked immunosorbent assay (ELISA) techniques. This too has not been used routinely by many forensic biology laboratories in the United States.

Era of Identification: Species Origin of Blood

Regardless of the controversy in the late 1800s and early 1900s over the ability of the catalytic tests to identify the presence of blood, the mere presence of blood on an object of important physical evidence does not dispel arguments regarding its origin. Valid reasons exist for individuals to have blood on their clothing or on weapons. Without confirmation that the blood is human, the test result is merely a piece of circumstantial evidence whose probative value will be argued by counsel. Identifying the blood as human takes the explanation of its presence to a higher plane; a person must have a good, verifiable reason to have human blood present to eliminate suspicion.

With the discovery that animal and human blood could be differentiated immunologically, this argument was dispelled. With the discovery of human-specific "antisera" in 1901,[35] a new forensic test had been born and a new era in forensic biology ushered in. The production and use of biological testing reagents consumed the attention of forensic biologists for the next 60-plus years.

Like the catalytic tests, the use of species-specific antisera for the determination of the presence of human or other animal blood had its proponents and antagonists.[36–44] Its application to forensic studies was rapid. Before long, however, problems in both the test's sensitivity and specificity were uncovered. The science needed a method of controlling the quality of the antisera being used. In 1901, P. Uhlenhuth first noted a variability in the production of antisera and suggested appropriate testing of the antisera prior to use in a forensic case.[45–46] This was probably the first recorded attempt at quality assurance related to forensic biological testing. Though quality assurance has become a major concern of modern forensic biologists, that it is not a new concept is illustrated in these early twentieth-century publications.

Initially, the species identification test—the precipitin or ring test—was performed in a tube; a positive reaction was the formation of a cloudy band or ring between the known anti–human serum and the unknown blood sample being tested. The specificity of the test became suspect if precautions, such as centrifugation of the bloodstain extract, were not taken in the preparation of the unknown specimen.[47]

Eventually, the test came to be performed in a gel matrix where the human-specific antisera and an unknown blood sample are permitted to come into contact. The test has two forms. One uses diffusion through the gel, a passive technique developed by Ouchterlony in 1949;[48] the other uses an electric current, the crossover electrophoresis technique described by Culliford[67] and by the Metropolitan Police Laboratory.[49] These latter modifications of the species test are more reliable and more sensitive than the original ring test.

Era of Differentiation: Finding Blood Groups

ABO: The First Human Polymorphism

The next era in forensic biology's history, the use of immune or naturally occurring sera to identify markers in blood that can serve to differentiate one blood sample from another, evolved from the same beginnings and at about the same time as the species test. The era of differentiation started with Landsteiner's 1900 discovery[50] that individuals could be classified into major groups, which he termed A, B, and C (now called the ABO blood group system). With this discovery, Landsteiner documented the first polymorphism in man. *Polymorphism*, literally "many forms", refers to the existence of two or more genetically determined alternative forms of a marker in a population with frequencies too great to be maintained by mutation alone.

Three nomenclature systems caused initial confusion in identifying blood group donors, in interpreting results of serological tests, and even in compatibility testing. One system, suggested by Jansky in 1907, designated blood groups by numbers (Roman numerals); Moss proposed a system that reversed these numbers. The situation was resolved when the A, B, O nomenclature of Landsteiner was made mandatory by the United States military in 1941.[51]

The ABO blood groups remained the only known polymorphism for 25 years. During this period, the markers were used to differentiate bloodstains using techniques promulgated mostly by Landsteiner[52] and Lattes (Lattes crust test).[53]

In the United States, the first forensic scientist to routinely identify the common blood groups from dried blood found on physical evidence was Alexander Weiner. Archival casework notes document that, as a staff bacteriologist in the New York City Medical Examiner's serology laboratory, he began using the absorption–inhibition test to identify blood groups on physical evidence in 1935. Discussions with his technician, Eve Gordon, documented that Weiner preferred the absorption–inhibition technique over the absorption–elution technique for the identification of the cell surface blood group antigens and used the Lattes test to identify the serum-based corresponding antibodies.

Although the first reported criminal case involving the differentiation of blood groups was reported by Lattes in 1916,[54] ABO typing on evidence of forensic importance in U.S. forensic biology laboratories did not become widespread until the 1960s. This coincided with a number of publications dealing with modifications of the absorption–elution technique, which was used extensively at the Metropolitan Police Laboratory in London.[55–56]

The absorption–elution technique has many advantages over the more widely used absorption–inhibition method, including increased sensitivity and the ability to use threads from fabric evidence directly. A positive absorption–elution test combined with Lattes crust test results constitutes a conclusive test for the blood group of unknown bloodstains. Unfortunately, the elution test is more sensitive than the Lattes crust test and the antigens being detected are more stable in evidentiary specimens than the naturally occurring antibodies identified in the Lattes crust test; this results in more positive results from the elution test. This is especially true with old stains because the red cell A and B antigens and H substance are more stable than the serum-based A and B antibodies.

MN: The Second Human Polymorphism

Other antigenic polymorphic systems were discovered and ultimately shown to have value for the analysis of forensic evidence. Among these, the

MN and Rh systems have been employed sparingly in American crime laboratories. The reasons are largely technical: it is difficult to interpret the results obtained.

For example, the MN system, described by Landsteiner and Levine in 1927,[57] was immediately recognized as another method of distinguishing blood samples having forensic significance. Problems with the system surfaced early on, however, when it was discovered that anti-N sera cross-react with M cells. Although the problem has a biochemical explanation, in practice the solution to the problem can lead to typing and interpretational errors.[58]

Rh: The Third Important Human Polymorphism

Before the discovery of the HLA serotypes and DNA tandem repeat polymorphisms, the Rh system was considered one of the more complex known. Identified by Landsteiner and Weiner in 1940 using immune serum prepared from Rhesus monkeys, it was the third polymorphism discovered in humans. It was also a system that had potential for use in forensic identifications.[59]Aside from its complex genetic considerations, which polarized advocates among Rh theorists, the system's other confusing aspects stemmed from three different nomenclature systems. The Fisher–Race system is the most widely used today,[60] but Weiner's Rh–Hr system had vociferous backers, especially in the New York area;[61] Murray's numerical system added to the confusion.[62]

The potential of the Rh system to differentiate blood is extremely good, but nomenclature aside, the complexity of the system requires an analyst to identify a complete phenotype in order to realize this potential. Unfortunately, a complete phenotype requires the identification of the absence of an antigen. Though it is possible to identify Rh antigens in dried blood,[63–64] assigning the nonexistence of an Rh antigen based on negative results is problematic; the inability to detect its presence could occur for other reasons, such as degradation, a weak antiserum, or small sample size. As always, there were differences of opinion concerning the use of Rh system in casework.[65] Fortunately, this problem was short-lived.

Although the system had been known since 1940, reliable detection using the absorption–elution test was not available to forensic biologists until the 1960s and 1970s. Unfortunately for Rh enthusiasts, this coincided with the successful application of electrophoretic methods for the identification of isoenzyme and protein polymorphisms in dried specimens. Thus, the problems in interpreting phenotypes from dried blood, the weakly reacting anti-Rh sera, the diverse and confusing nomenclature systems, and the emergence of the isoenzyme polymorphisms as a contemporary identification alternative method probably limited its use in American forensic biology laboratories. Some laboratories, of course, did use the system routinely.

Era of Individualization, Phase I: Before DNA Profiling

Differentiating one blood sample from another served the needs of forensic biologists for almost 70 years. The discovery of several genetic marker systems that are stable in dried blood and the successful adaptation of technology to identify these markers planted the seed of a notion that a blood sample could be identified as coming from one person to the statistical exclusion of all others. This concept was termed *individualization*, and, though it was never a reality, at least in its first phase, results were obtainable from forensic specimens that drastically eliminated the bulk of the population from consideration as possible donors.

Isoenzymes and Proteins: Electrophoretic Methods

After 1965, the electrophoretic separation of soluble isoenzymes and proteins in a gel matrix to identify enzyme and protein polymorphisms became the hot technique for forensic biologists. The simplicity of the techniques and relatively low start-up costs were instrumental in expanding the technology to virtually all forensic biology laboratories in the United States.

1967–1977: The Starch Years. The credit for introducing this biochemical approach of individualization belongs to the Metropolitan Police Laboratory in London. It recognized that these markers would be valuable forensic tools and devised a thin-layer starch gel technique that maximized sensitivity and enzyme activity during prolonged electrophoresis at higher voltages.[66] In the United States, the Pittsburgh and Allegheny County Crime Laboratory, on-line with phosphoglucomutase (PGM_1) in 1969, was one of the first laboratories to use the technology in criminal cases. Under sponsorship of the Law Enforcement Assistance Administration (LEAA), in 1971 Culliford published a monograph describing in detail the practical use of electrophoretic techniques for the identification of isoenzyme markers in dried blood;[67] this became the first definitive procedures manual used by U.S. forensic biology laboratories.

The number of reliably identifiable isoenzyme and protein polymorphisms in dried blood rapidly expanded during the 1970s, and by the early 1980s many laboratories boasted of having the technical capability to identify 15 or more polymorphic systems from dried blood. Most laboratories chose to routinely analyze for the handful of markers with good discrimination potential. They included PGM_1,[66,68–70] adenylate kinase (AK),[71] erythrocyte acid phosphatase (ACP_1 or EAP),[72–74] and esterase D (EsD).[75–76] Normally, each marker was analyzed as a separate system.

1977–1986: "Starch Wars" and the Multisystem. The late 1960s and mid-1970s can be characterized as the starch years because most of the isoenzyme polymorphisms were identified using adaptations of the original horizontal, thin-layer starch gel technique described by Culliford.[67] In the mid-1970s, Grunbaum published alternative techniques for identifying these markers using cellulose acetate membranes as the separation medium,[69] methods that were adopted in a few forensic biology laboratories. Grunbaum's techniques had advantages over the starch gel methods: they were faster, could use smaller sample sizes, and required no cooling. However, the cellulose acetate methods arguably do not have the resolving power of the starch techniques, so some forensic biologists were confused as to which was best to use.

In an effort to determine which electrophoretic matrix (starch gel, cellulose acetate, agarose, acrylamide, some other, or a hybrid separating medium) was the best of the state-of-the-art methods, the National Institute of Justice, through LEAA, awarded a contract to Aerospace Corporation that dealt with the issue of genetic marker identification and included as a part of its statement of work, "the development of improved methods for the electrophoretic identification of blood genetic markers."[77] Outside proposals were solicited; one submitted by Beckman Instruments was approved. Beckman Instruments subcontracted to Grunbaum, the developer of the cellulose acetate systems, at the University of California. The proposal brought together Grunbaum's laboratory: Brian Wraxall, one of the developers of the original thin-layer starch gel system; two research associates to support Grunbaum's efforts; and M. D. Stolorow, an expert in the electrophoretic analysis of blood stains. I was the project manager for Aerospace Corporation and liaison with LEAA. The project was initiated in December 1976 and completed in June 1978.

The final product was the Bloodstain Analysis System (BAS), commonly known as the multisystem, which permitted the identification of multiple markers simultaneously; the markers were analyzed in three separate groups. Group I permitted the identification of PGM_1, EsD, and glyoxalase (GloI) on an agarose gel containing 2% starch (added to permit the visualization of the GloI isoenzymes).[78] Group II permitted the identification of ACP_1, AK, and adenosine deaminase (ADA) on a thin-layer horizontal starch gel. Group III permitted the identification of group specific component (Gc) and haptoglobin (Hp) from a single bloodstain extract, each system being analyzed on a separate electrophoresis system. By the mid-1980s, these systems were being used by the majority of forensic biology laboratories in the United States.

Unfortunately, the research did not proceed smoothly. There were differences of opinion regarding the direction of the project and, after nine months, Grunbaum resigned. The remainder of the work was completed at Beckman Instruments in Anaheim, California. The participants began to affectionately call the project "Starch Wars" in reference to the starch gel versus cellulose acetate controversy that began in the mid-1970s and escalated during their project's research and development phase.

The project culminated in the training of four local forensic biology laboratories (the Los Angeles County Sheriff's Office, the Georgia Bureau of Investigation, the Minnesota State Police, and the New Jersey State Police) and the FBI in the newly developed methods. These laboratories participated in demonstration and proficiency testing over a three-month period.[77] A follow-on LEAA contract was awarded to the Serological Research Institute (SERI) to train American forensic biologists in the new techniques.

1978–1986: Frye Battles. The Bloodstain Analysis System, or multisystem, was the focus of numerous legal battles in the 1980s; these became venomous personal feuds between forensic scientists.[79–81] In hindsight, these forays were but skirmishes that were blueprints for the future DNA battles; virtually all of the scientific participants in the original LEAA/Aerospace/Beckman research and management teams became involved, but only a few nonforensic scientists took part.

Although these battles wreaked havoc with the personal and sometimes private lives of the participants, the eventual winner, measured in terms of laboratory quality, was the criminal justice system. These hearings addressed questions raised by the Frye admissibility ruling (see "The Demise of the Frye Rule" section in Chapter 9): the relevant scientific community, the reliability of the methods, the record-keeping practices of the laboratories performing the tests, the educational backgrounds of the laboratory personnel, and the meaning of forensic scientist versus forensic technician. In most jurisdictions, the reliability of the methods was upheld.

Better Discrimination with Isoelectric Focusing. Inevitably, Starch Wars had to make room for newer, more discriminating technology. This new technology, isoelectric focusing (IEF), provides separations of isoenzyme and protein polymorphisms in pH gradient; it was introduced to forensic biologists in the late 1970s and widely adopted in the 1980s. Much of this original work was done at the Home Office in England and at the FBI in Quantico, Virginia. But though IEF provided laboratories with a better method of resolving genetic markers, it did not completely do away with the multisystem.[82–89]

Era of Individualization, Phase II: After DNA Profiling

Today, scientists, amidst their personal biases and agendas, are searching to establish the truth about the reliability and ultimate value of DNA profiling. Through all this, DNA profiling is being admitted in the courts as a tool to help decide alleged facts in criminal and civil cases throughout the world. This has been a highly publicized endeavor, with special interest groups striving to impose their agendas in any way possible. As discussed earlier, the controversy surrounding DNA profiling is not unique in the history of forensic biology. Even the simplest (by today's standards) of scientific tests used to identify the presence of blood were plagued with scientific controversy as they were offered to the courts.

1986 to Present: Enter DNA

Although DNA profiling was introduced to law enforcement in the United States in 1986 by Lifecodes Corporation and Forensic Science Associates, 1987 marks the most dramatic entry of DNA to the legal system through the conviction of Tommy Lee Andrews in a Florida rape case.[90] Andrews, a former pharmaceutical clerk, was convicted of attacking an Orlando woman in her home; she could not identify her attacker, but semen from a vaginal swab had DNA that matched Andrews's.

By 1990, forensic serologists became forensic biologists. Under sponsorship of the FBI, local crime laboratories, academics, the federal government, and interested private parties formed the Technical Working Group for DNA Analysis Methods, which formulates recommendations for quality guidelines for forensic DNA typing laboratories.[91] For the first time, the defense bar mobilized, academicians with no background or prior interest in forensic science became adversaries, legislators found political fodder on which to bolster their reputations, and legislation was proposed to have DNA profiles performed on convicted felons. The federal government sponsored three separate reports on forensic DNA testing. The first was Congress's Office of Technology Assessment report and the second two were National Research Council reports sponsored by the National Institute of Justice.[92,93, 95] Again, experts and egos battled as the science has began its arduous path toward acceptance.

Much of the credit for increased awareness of the laboratory's responsibility in performing quality testing is due to the 1989 Castro decision.[94] This decision may have changed the Frye approach and made laboratory practice and reliability an integral part of the admissibility criteria. The case, in which a pregnant woman had been brutally murdered in her Bronx apartment, had the unusual twist of having the experts meet outside the courtroom. The judge evoked a three-pronged test. First, is there a theory behind DNA technology that is agreed upon? Second, can the technology be used reliably for the analysis of forensic evidence? Third, did the laboratory perform the tests properly so that the results are reliable? The judge ruled that the laboratory, Lifecodes Corporation, did not perform the test using the necessary quality controls to ensure reliability. The Castro court decision is important because it established the notion that quality assurance would become an important parameter for the admissibility of DNA evidence.

Conclusion

"What is past is prologue." Forensic tests have progressively become more important to the outcome of criminal and civil proceedings. As this trend continues, more rigorous challenges can be expected and standards and controls will be mandated in order to provide confidence that forensic laboratories are employing appropriate procedures and that the scientists are performing the tests reliably. As new technologies are applied to forensic biological problems, the lessons of the past should not be forgotten.

References

1. *State of New York vs. Lamont Barner*. Ind. No. 05270/92.
2. Van Deen, J. *Arch. Hollaend. Beitr. Natur-Heilk* **1862,** *3(2)*, 228–231.

3. Taylor, A. S. *Guy's Hosp. Rep.* **1868,** *13(3)*, 431–455.
4. Jenne, J. N. *Vermont Med. Monthly* **1896,** *2(5)*, 127–137.
5. Siefert, E. *Gerichtl. Med. Oeff. Sanitaetswes* **1898,** *16(3F)*, 1–27.
6. Chapman, H. C. *A Manual of Medical Jurisprudence and Toxicology;* W. B. Saunders: Philadelphia, PA, 1892.
7. Delarde, F.; Benoit, A. C. R. *Soc. Biol.* **1908,** *64*, 990–992.
8. Delarde, F.; Benoit, A. C. R. *Soc. Biol.* **1908,** *64*, 1048–1050.
9. Liman, H. *Vierteljahrschr. Gerichtl. Med. Oeff. Sanitaetswes* **1863,** *24*, 193–218.
10. Mecke, A.; Wimmer, C. *Chem. News (London)* **1895,** *71(1851)*, 238.
11. Whitney, W. F. *Boston Med. Surg. J.* **1909,** *160*, 202–203.
12. Alsberg, C. L. *Arch. Exp. Pathol. Pharmakol., Suppl.* **1908,** 39–53.
13. Dervieux, F. *Arch. Int. Med.* **1911,** *Leg 1 (suppl)*, 92–94.
14. Camps, R. E.; Robinson, A. E.; Lucas, B. G. B. *Gradwohl's Legal Medicine,* 3rd ed.; John Wright & Sons: Bristol, England, 1954.
15. Kastle, J. H. *Chemical Tests for Blood;* U.S. Hygienic Laboratory Bull. No. 51; U.S. Public Health and Marine Hospital Service. U.S. Government Printing Office: Washington, DC, 1909.
16. Adler, O.; Adleer, R. *Hoppe-Seyler's Z. Physiol. Chem.* **1904,** *41*, 59–67.
17. Garner, D.; Cano, K. M.; Peimer, G. S.; Yeshion, T. E. *J. For. Sci.* **1976,** *21*, 816–821.
18. Utz, H. *Chem-Zig.* **1903,** *27(94)*, 1151–1152.
19. Kastle, J. H.; Shedd, O. M. *Am. Chem. J.* **1901,** *26*, 526–539.
20. Meyer, E. *Beitrage zur Leukocytenfrage. Muench. Med. Wochenschr.* **1903,** *50(35)*, 1489–1493.
21. Specht, W. *Angew. Chem.* **1937,** *50*, 155–157.
22. Specht, W. *Dtsch. Z. Gesamte Gerichtl. Med.* **1937,** *28*, 225–234.
23. Grodsky, M.; Wright, K.; Kirk, P. L. *J. Crim. Law Criminol. Police Sci.* **1951,** *42*, 95–104.
24. Sutherland, W. D. *Bloodstains: Their detection and determination of their source;* Balliere, Tindall & Cox: London, 1907.
25. Kerr, D. *Br. Med. J.* **1926,** *1*, 262.
26. Dilling, W. *Br. Med. J.* **1926,** *1*, 134.
27. Teichmann, L. A. *Ration. Med.* **1853,** *3*, 375.
28. Beam, W.; Freak, G. A. *Biochem. J.* **1915,** *9*, 161.
29. Wagenaar, M. Z. *Anal. Chem.* **1935,** *103*, 417.
30. Takayama, M. *Kokka Igakkai zasshi* **1912,** *306*, 15.
31. Lee, H. C. *Forensic Science Handbook;* Saferstein, R., Ed.; Prentice-Hall: England Cliffs, NJ, 1982, pp 279–280.
32. Grunbaum, B. W. *Beckman Electrophoresis Information* **1976,** *3(3)*.
33. Fiori, A. *J. Forensic Sci.* **1961,** *6*, 459.
34. Lee, H. C.; DeForest, P. *The use of anti-human Hb serum for bloodstain identification;* 29th Annual Meeting, Amer. Acad. Forensic Sci.: San Diego, CA, 1977.
35. Uhlenhuth, P. *Dtsch. Med. Wochenschr* **1901,** *27(6)*, 82–83.
36. Ziemke, E. *Vierteljahrschr. Gerichtl. Med. Oeff. Sanit.* **1901,** *22(3F)*, 231–234.
37. Frenkel, H. *Arch. Anthropol. Crim. Med. Leg.* **1901,** *16*, 649–654.
38. Stern, R. *Dtsch. Med. Wochenschr.* **1901,** *27(9)*, 135.
39. Corin, G. *Arch. Anthropol. Crim. Med. Leg.* **1901,** *16*, 409–419.
40. Wood, E. S. *Med. Leg. J.* **1903,** *21*, 26–32.
41. Marshall, H. T.; Teague, O. *Philippin J. Sci.* **1908,** *3B*, 357–377.

42. Graham-Smith, G. S.; Sanger, F. *J. Hyg.* **1903,** *3,* 258–291.
43. Kratter, J. *Arch. Krim.-Anthropol. Kriminalistik* **1902,** *10,* 199–209.
44. Schulze, O. *Med.-Leg. J.* **1903,** *21,* 39–40.
45. Uhlenhuth, P. *Dtsch. Med. Worchenschr.* **1901,** *27(30),* 499–501.
46. Schleyer, F. In *Methods of Forensic Science;* Lundquist, F., Ed.; Interscience: New York and London, 1962; Vol. 1, pp 291–333.
47. Lee, H. C. In *Forensic Science Handbook;* Saferstein, R., Ed.; Prentice-Hall: England Cliffs, NJ, 1982, pp 267–337.
48. Ouchterlony, O. *Acta Pathol. Microbiol. Scand.* **1949,** *26,* 507.
49. *Biology Methods Manual;* Metropolitan Police Forensic Science Laboratory: 1978; pp 2-94–2-99.
50. Landsteiner, K. *Centralbl. Bakteriol. Parasitenkd. Infektionskr.* **1900,** Abstr. I 27, 357–362.
51. Camp, F.; Conte, N.; Ellis, F. *Military Blood Banking: Genetics for the Reference and Forensic Testing Laboratory.* A Monograph. U.S. Army Medical Research and Development Command: Washington, DC, 1971; pp 16–17.
52. Landsteiner, K; Richter, M. Z. *Medizinalbeamte* **1903,** *16(3),* 85–89.
53. Lattes, L. *Arch. Antropol. Crim. Psichiatr. Med. Leg.* **1913,** *34(4 ser. 5),* 310–325.
54. Lattes, L. *Arch. Antropol. Crim. Psichiatr. Med. Leg.* **1916,** *37(4 ser. 7),* 298–303.
55. Kind, S. S. *Nature (London)* **1960,** *185,* 397–398.
56. Kind, S. S. *Nature (London)* **1960,** *187,* 789–790.
57. Landsteiner, K.; Levine, P. *Proc. Soc. Exp. Biol. Med.* **1927**, 24, 600–602.
58. Shaler, R. C.; Hagins, A. M.; Mortimer, C. E. *J. For. Sci.* **1978,** *23,* 570–576.
59. Landsteiner, K; Weiner, A. S. *Proc. Soc. Exp. Biol. Med.* **1940,** *43,* 223.
60. Fisher, R. A.; Race, R. R. *Nature (London)* **1946,** *157,* 81–89.
61. Weiner, A. S. *Science (Washington, DC)* **1945,** *102,* 479–482.
62. Murray, J. *Nature (London)* **1944,** *154,* 701–702.
63. Nickolls, L. C.; Pereira, M. *Med. Sci. Law* **1962,** *2,* 172–179.
64. Bargagna, M.; Pereira, M. *J. For. Sci. Soc.* **1967,** *7,* 123–130.
65. Martin, P. D. *J. For. Sci. Soc.* **1977,** *17,* 139–142.
66. Wraxall, B. G. D.; Culliford, B. J. *J. For. Sci. Soc.* **1968,** 8, 81–82.
67. Culliford, B. *The Examination and Typing of Bloodstains in the Crime Laboratory;* U.S. Government Printing Office: Washington, DC, 1971.
68. Culliford, B. *J. For. Sci. Soc.* **1967,** *7(3),* 131–133.
69. Grunbaum, B. *J. For. Sci. Soc.* **1974,** *14,* 151–157.
70. Zajac, P.; Sprague, A. *J. For. Sci. Soc.* **1975,** *15,* 69–74.
71. Culliford, B.; Wraxall, B. G. D. *J. For. Sci. Soc.* **1968,** 9, 79–80.
72. Wraxall, B. G. D.; Emes, E. *J. For. Sci. Soc.* **1976,** *16,* 127–132.
73. Sensabaugh, G.; Golden, V. Erythrocyte acid phosphatase typing on starch gels containing glycerol. *For. Serology Newsletter* **1976** *2(6).*
74. Grunbaum, B.; Zajac, P. *J. For. Sci.* **1978,** *23(1),* 84–88.
75. Parkin, B.; Adams, E. *Med. Sci. and the Law* **1975,** *15(2),* 102–105.
76. Grunbaum, G.; Harmor, G.; Del Re, B.; Zajac, P. *J. For. Sci.* **1978,** *23(1),* 89–93.
77. Wraxall, B. G. D.; Bordeaux, J.; Harmor, G.; Walsh, J. *Final Report: Bloodstain Analysis System;* Beckman Advanced Technology Operations, 1978.
78. Wraxall, B. G. D.; Stolorow, M. D. *J. For. Sci.* **1986,** *31(4),* 1439–1449.
79 *State of Kansas v. Adrian C. Washington,* 229 Kan. 47, 622p. 2d 986 (1981).
80. *State of Michigan v. Jeffrey Allen Young,* 418 Mich.1, 340 N.W. 2nd 805 (Mich. 1983).

81. *State of California v. Brown,* 40 Cal. 3d. 512, 1985.
82. Baxter, M.; Randall, T. W.; Thorpe, J. W. *J. For. Sci. Soc.* **1982,** *22(4),* 367–372.
83. Budowle, B. *Electrophoresis* **1984,** *5(3),*165–167.
84. Budowle, B. *Electrophoresis* **1984,** *5(4),* 254–255.
85. Budowle, B. *Electrophoresis* **1984,** *5(5),* 314–316.
86. Budowle, B. *Electrophoresis* **1985,** *6(2),* 97–99.
87. Divall, G. B. *For. Sci. Int.* **1985,** *28,* 277–285.
88. Divall, G. B.; Ismail, M. *For. Sci. Int.* **1983,** *22,* 253–263.
89. Murch, R. S.; Budowle, B. *J. For. Sci.* **1986,** *31(3),* 869–880.
90. *State of Florida v. Tommy Lee Andrews,* 533 So. 12d 841 (D.C.A. Fla. 1989).
91. Mudd, J. *Crime Laboratory Digest* **1989,** *16(2),* 40–59.
92. *Genetic Witness: Forensic Uses of DNA Tests;* U.S. Congress, Office of Technology Assessment. U.S. Government Printing Office: Washington, DC, 1990; OTA-BA-438.
93. National Research Council, Committee on DNA Technology in Forensic Science. *DNA Technology in Forensic Science;* National Academy Press: Washington, DC, 1992.
94. *People v. Castro,* 144 Misc. 2d 956, 545 N.Y.S. 2d 985 (Sup. Ct. Bronx Co. 1989).
95. National Research Council. *The Evaluation of Forensic DNA Evidence;* National Academy Press: Washington, DC, 1996.

3

DNA Typing Methods for Forensic Analyses

David H. Bing

Deoxyribonucleic acid (DNA) is found in every living cell that replicates. Encoded in its structure is all genetic information that forms the basis for life. DNA is a linear molecule consisting of two strands of four nucleic acids (adenosine, thymidine, cytosine, guanosine) that are covalently linked by phosphodiester bonds. These two strands interact noncovalently to form an alpha-helical structure. During cell division, the strands unwind and via an enzymatic process a complementary copy of each strand is produced. Each of the two new strands reassociate and are partitioned as a double-stranded alpha-helical molecule to one of the two cells. In this way, all genetic information is faithfully and completely transferred to every cell in the organism. Since the 1944 discovery by Avery, MacCleod, and McCarty that DNA is the principal source of genetic material in a cell,[1] chemists, physicists, and biologists have developed a vast array of methodologies and techniques to investigate and elucidate the relationship between the structure and function of DNA.

The application of this DNA technology to the analysis of forensic evidence began in the early 1980s, about 10 years after molecular biology became a recognized scientific discipline. Forensic analyses based on DNA have focused on molecular diagnostic techniques that allow for analysis of polymorphic genetic markers encoded in the linear sequence of the DNA. In most instances, the techniques that have been applied to forensic analyses are those developed first to study the biophysical and biological properties of DNA. Only within the past few years have DNA-based methods emerged that were specifically designed and optimized for the analysis of the trace biological materials that often constitute the majority of the evidence collected at crime scenes. This chapter reviews the technological innovations and discoveries in molecular biology during the past 20 years that have found immediate applications in forensic DNA analysis, describes the newer, specific forensic DNA methods, and synopsizes what has happened when this technology has been used in the criminal justice system for both exculpatory and inculpatory purposes.

Reason for Using DNA for Forensic Testing

The laboratorian must demonstrate two things before being able to successfully apply scientific methodology to forensic diagnostic testing of physical evidence collected at crime scenes: that the sample is appropriate for the specific analysis to be used and that the technology of the test is based on sound scientific principles. G. Sensabaugh elaborated on these two principles to describe six desirable aspects of a marker to be used in a diagnostic test in the analysis of physical evidence collected at crime scenes.[2]

1. The marker polymorphism should involve simple qualitative differences such that an individual can be assigned to one type or another without ambiguity. It is desirable but not necessary that the phenotype directly reflect the genotype.
2. The marker should be expressed in a tissue of interest (e.g., blood, semen) throughout the life of the individual; markers expressed only during restricted periods of development (e.g., fetal proteins) or only in inaccessible tissues (e.g., brain) are of limited value.
3. The marker should be robust; that is, it should be able to survive drying and other travails typically experienced by biological evidence. In particular, the marker cannot undergo changes that make one type look like another or that so confuse the typing analysis that the risk of mistyping becomes significant.
4. The analytical procedures for marker typing should be relatively simple and straightforward.

5. The analytical procedure should consume as little as possible of the sample being tested; the less sample consumed in any single test, the more tests that can be done.

6. Finally, the polymorphism should subdivide the population fairly evenly; a polymorphism that splits the population 50:50 is of greater value than one with a 99:1 split.

The physical and biological properties of DNA almost completely meet all these criteria. The properties and structure of DNA have been described and summarized.[3] The following properties are relevant to the analysis of forensic evidence:

- DNA contains the genetic code for all living cells; with the exception of identical twins, it is believed that the DNA of each individual is different. Thus, the DNA of every human has the potential to be used as a unique identifier of that person.[4]

- Every cell in a living organism contains a complete copy of the DNA of the individual from whom it is obtained. Thus, any cellular material from an individual potentially yields the same identifier, making it possible to compare disparate types of biological samples, such as blood, semen, and saliva.[3]

- DNA is a robust molecule. It comprises two strands of two purine nucleotide bases (deoxycytosine triphosphate and deoxyguanosine triphosphate) and two pyrimidine nucleotide bases (deoxyadenosine triphosphate and deoxythymidine triphosphate) that covalently link by phosphodiester bonds in a linear array (see Figure 1). These two strands are bound together in a double alpha-helix that is able to survive physical denaturation as well as being relatively resistant to natural catabolic processes. Thus, any cellular sample that has been stored under conditions of maintenance favorable to the preservation of the structure of DNA can potentially be analyzed successfully. For example, DNA isolated from mummies as well as fossils has been successfully analyzed for the presence of specific genetic markers.[5–7]

- The analytical sensitivity achievable based on analysis of DNA is high. Current molecular biology methods allow for analysis of a sample that contains as few as 100 molecules of DNA, meaning that for markers where there is a single copy of the gene per cell, as few as 100 cells need be analyzed.[8]

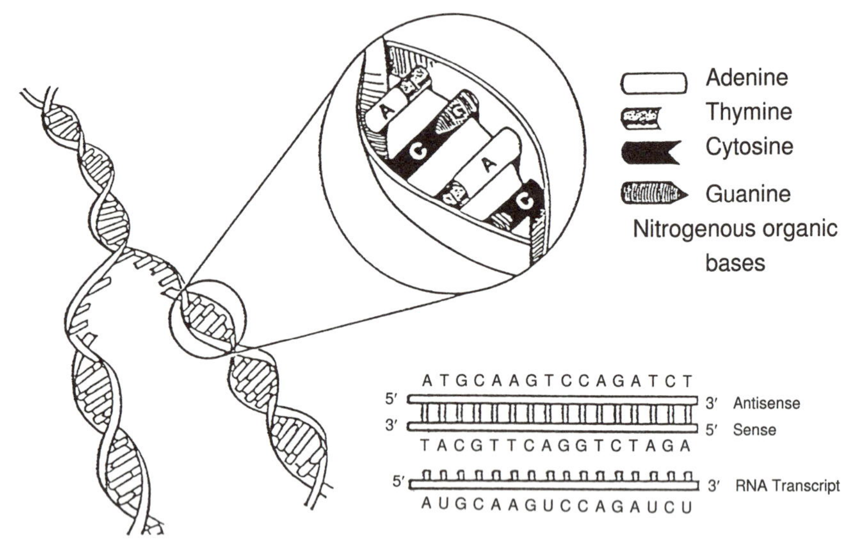

Figure 1. The double-stranded structure of DNA undergoing semiconservative replication. Each parental strand is a template for the synthesis of a new daughter strand formed according to the base pairing rules: A=T, C=G. The new doublehelix molecules are identical, each consisting of an original parental and a new daughter strand. The sense (3′-5′) and antisense (5′-3′) designation of the molecule for mRNA transcript production is also illustrated. Reproduced with permission from reference 3. Copyright 1992 by W. H. Freeman and Company.

- DNA contains the genetic code; every cell contains a complete copy of this code. Thus, all polymorphic genetic systems are potentially suitable targets for genetic typing in a forensic sample, as long as the forensic sample contains cellular material or DNA molecules.[3–4]

These properties of DNA allow for direct analysis of genetic markers encoded in the linear sequence of DNA and form the basis for the molecular analysis of forensic biological materials.

DNA Technologies First Used in Forensic Testing

Forensic DNA testing initially combined three tools developed for studying human genetic systems by direct analysis of DNA: Southern blotting, a method that allows for separation of DNA molecules into patterns that reflect the molecular structure of DNA;[9–10] restriction fragment length polymorphism, a type of genetic analysis originally described as a method of making a linkage map of the entire human genome;[11] and variable number tandem repeat markers, highly polymorphic genetic markers widely distributed on all human chromosomes.[12]

Southern Blotting

Southern blotting is a technique developed by E. M. Southern in 1975 to directly transfer fragments of DNA to solid support, such as a nylon membrane, following their separation by electrophoresis in semisolid media such as agarose gels. The transfer occurs by capillary action of salt solutions. Once transferred, the DNA is irreversibly bound to the membrane either by heating the membrane or exposing it to ultraviolet light. This process results in a print containing the transferred DNA fragments on the membrane, which in turn can be analyzed for specific genetic markers by hybridization to characterized sequences of DNA, called DNA probes, in which the sequence of bases code for the marker.[9] A further refinement by Elder and Southern was their development of a method for characterizing the fragments of DNA transferred by blotting and detected by the DNA probes based on molecular weight of the DNA fragments.[10] Southern blotting was the first technique used in DNA forensic analysis and continues to be one of the fundamental methods used by forensic laboratories doing DNA analysis.

Restriction Fragment Length Polymorphism

Restriction fragment length polymorphism (RFLP), described in 1985 by D. Botstein, R. L. White, M. Skolnick, and R. W. Davis, is a type of genetic analysis that has allowed for construction of a genetic linkage map of the human genome.[11] The method uses DNA probes that recognize sequence polymorphisms of single-copy genes when the probe is hybridized to DNA digested with restriction enzymes. These enzymes are DNA hydrolases that cleave double-stranded DNA internally at specific palindromic base recognition sequences. In classical RFLP analysis, the location of the restriction site is polymorphic, which leads to generation of different length fragments of DNA. That is, the different length DNA fragments are generated because the distance between two fixed recognition sites for the restriction endonuclease is genetically variable. When a variable restriction site is present between two fixed recognition sites for the restriction enzyme, two fragments are produced; if not, only one fragment is formed. An individual may be homozygous or heterozygous for the variable restriction site. Thus, when the site is absent or present on both chromosomes, it is homozygous; it is heterozygous when absent on one chromosome and present on the other. This site is inherited in Mendelian fashion. The basis of the change in fragment length as well as the recognition site is believed to be the insertion or deletion of nucleotides. RFLP analysis was the first DNA-based test used by testing laboratories for the analysis of forensic evidence and, in 1987, played a part in the first criminal case that used DNA for analysis of forensic evidence in the United States.[12]

Variable Number Tandem Repeat

Variable number tandem repeat (VNTR) markers are a type of RFLP polymorphic genetic marker first reported by A. R. Wyman and R. L. White in 1980.[13] The polymorphism is based on the variation in length of tandem repeats of sections of DNA where there is an end-to-end duplication of a sequence of the bases composing the DNA. The size of the repeat may be as short as two or as long as several thousand bases.[14] As the name implies, the number of repeats varies and that number is the genetic marker (see Figure 2). Following digestion of genomic DNA with restriction enzymes, these markers can be detected by Southern blotting and hybridization to DNA-specific probes.[3]

Two types of VNTR markers have been used in DNA forensic analysis. In 1985, A. Jeffreys, V. Wilson, and S. L. Thein described hypervariable minisatellite regions that consist of a common core sequence repeated many times. The complexity of the RFLP pattern with this type of VNTR is due to the presence of restriction sites within the core sequence (Figure 2A), resulting in fragments of different lengths being generated from a single DNA sequence. Because of the many fragments produced, these markers are called *multilocus*. This type of marker has been used extensively in the United Kingdom for forensic testing.[15–20] The example first characterized by Jeffreys and co-workers was a sequence of 33 base pairs repeated four times in tandem within the human myoglobin gene.[15] The population genetics of this genetic marker system was detailed in a report from Jeffreyss laboratory in 1991[19].

The other kind of VNTR marker used for forensic testing also consists of repeated sequences of DNA, but the restriction enzyme acts at a site on either side of the repeated sequence (Figure 2B). This type of marker behaves like the kind of RFLP marker described by Botstein et al.,[11] in that

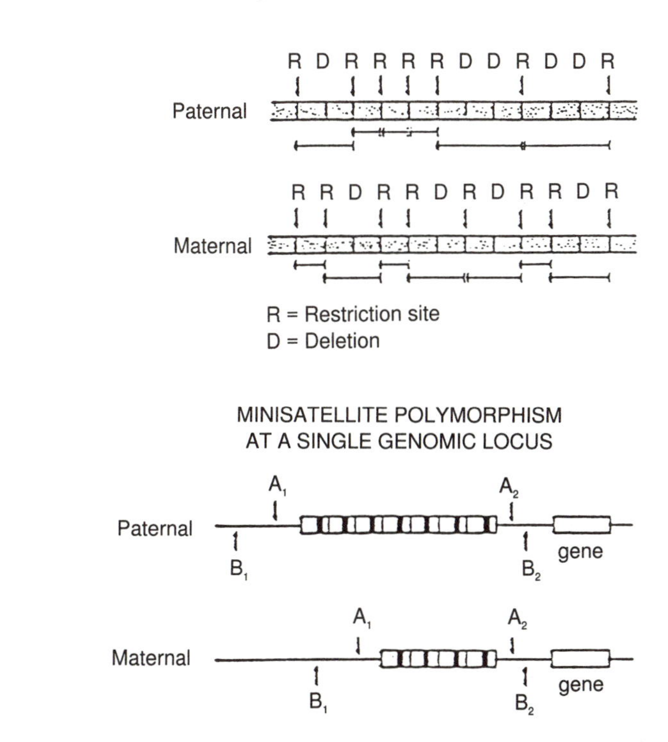

Figure 2. Multilocus and single-locus VNTR polymorphisms. A. Multilocus polymorphism: The origin of the multiple bands occurs because the restriction sites are within the tandem repeat sequences. B. Single-locus polymorphism: The origin of single bands occurs because restriction sites (A and B) are outside the tandem repeat sequences.

an individual has either one or two alternative forms of the marker, resulting in a single or double banded pattern. Single-locus VNTR markers, used by the Lifecodes Company in 1987 in the first criminal case based on DNA evidence in the United States, were the types of probes standardized and introduced by the FBI in late 1988[21] and that continue to form the basis for RFLP analyses now carried out by most U.S. forensic laboratories.[12]

The genetic properties of single-locus markers have been investigated and analyzed extensively since 1988. Single-locus VNTR markers appear to behave as a quasicontinuous distribution, with as many as 30 different markers possible.[22–25] These markers are inherited in a Mendelian fashion and, though there are different frequencies for different markers in different ethnic groups, all populations analyzed to date seem to have the same range of sizes for a given VNTR marker.[26–28] Concerning whether VNTR markers meet expectations of Hardy–Weinberg equilibrium and the requirements for linkage analysis, ambiguities arose because of the technical problems of measuring fragments of DNA that differed by fewer than 50 base pairs, because of the possible existence of null alleles and because of concerns about the size and randomness of human database samples used to generate estimates of the frequencies of the VNTR markers.[29–32] The results of the majority of analyses of population genetics of the VNTR markers and VNTR phenotypes observed were found to be consistent with the thesis that the VNTR markers usd in forensic analyses are Hardy–Weinberg equilibrium and do not exhibit linkage.[33–38]

The controversies regarding the use of VNTR markers for DNA-based forensic testing were addressed in a report, *DNA Technology in Forensic Science*, published in 1992 by the National Research Council.[39] In 1996, a second report, *The Evaluation of Forensic DNA Evidence*, was issued by the same organization in response to major criticisms of the first report.[40] A series of recommendations that were made in the 1996 report took into account recommendations in and criticisms of the 1992 report and included the following items:

- recommendations for proficiency testing to improve laboratory performance;
- the need for formal accreditation of forensic DNA laboratories;
- procedures for collecting and handling crime scene evidence;
- methods for estimating random match probabilities, determining allele frequencies, calculating the frequencies of combined genetic profiles, and modifying the frequencies to correct for possible differences due to substructure within populations;
- methods to search and use population databases containing DNA genetic profiles of convicted felons;
- the use of floating and fixed bins for estimating VNTR allele frequencies.

Prior to publication of the second report in 1996, Bruce Budowle and Eric Lander co-authored a 1994 editorial in *Nature*; they reviewed many of the controversies about forensic DNA-based genetic testing and concluded that "the DNA wars are over".[41] A review of the 1996 National Research

Council report by Bruce Weir, a population geneticist who severely criticized the 1992 report, echoed the opinions of Budowle and Lander. Thus, it appears that after six years of heated debate, a working consensus is beginning to form that DNA genetic testing based on VNTR markers is a scientifically acceptable application of human genetic testing.[42]

Two properties of single-locus VNTR markers make them attractive for forensic analysis. First, each of them represents a different genetic system. As any given sample of genomic DNA contains all known VNTR markers, it is possible to perform multiple analyses on one sample. Furthermore, frequencies of occurrence of each marker can, in principle, be combined to provide a unique pattern for an individual or a sample of DNA derived from a piece of forensic evidence if the VNTR markers used in the analysis segregate independently—that is, on separate chromosomes. This is the basis of the product rule introduced in 1986 by Lifecodes and in 1987 by the FBI to combine frequencies of each VNTR marker used in an analysis to make an estimate of the overall frequency of occurrence of the RFLP banding patterns obtained in the analysis of forensic evidence.[12] Second, the degree of heterozygosity of VNTR markers has been reported to be as high as 98%, thus lowering the probability that two unrelated individuals might share the same pattern of VNTR markers.[25,43–44] Recently, methods based on the polymerase chain reaction, described in the next section, have been used to detect VNTRs called short tandem repeats (STRs) where the repeat is 3 to 4 bases, as well as those where the repeat is between 7 and 250 bases.[45]

DNA Methods for Forensic Analysis Based on Polymerase Chain Reaction

One of the drawbacks to RFLP analysis with VNTRs has been the lack of sensitivity and stringent sample requirements. Although 50 ng of DNA can ideally be detected in a RFLP band, between 1 and 2.5 μg of high-molecular-weight (e.g., nonhydrolyzed) DNA is required to obtain a good RFLP pattern. This usually requires about 50 to 100 μL of blood or semen or at least 15 hairs with intact roots. Frequently, forensic samples contain 2 ng or less and, due to environmental factors such as bacterial contamination, the DNA obtained may have been hydrolyzed by nucleases that are part of all living cells. An alternative method that allows for analysis of small amounts of degraded human DNA was described in 1986 by R. K. Saiki, T. L. Bugawan, G. T. Horn, K. B. Mullis, and H. A. Erlich.[46] It is an in vitro method that uses a DNA polymerase to enzymatically and repetitively produce in a cyclical fashion copies of beta globin and DQ alpha alleles from human genomic DNA isolated from blood. The term *polymerase chain reaction* (PCR) is used to describe this system.

Briefly, the PCR process works as follows. The sections of DNA to be replicated are selected by synthetic DNA oligonucleotides called primers that are complementary to target DNA sequences located on opposite strands of the DNA but separated by intervening DNA sequences (see Figure 3). The DNA is converted into single strands by heat denaturation, and

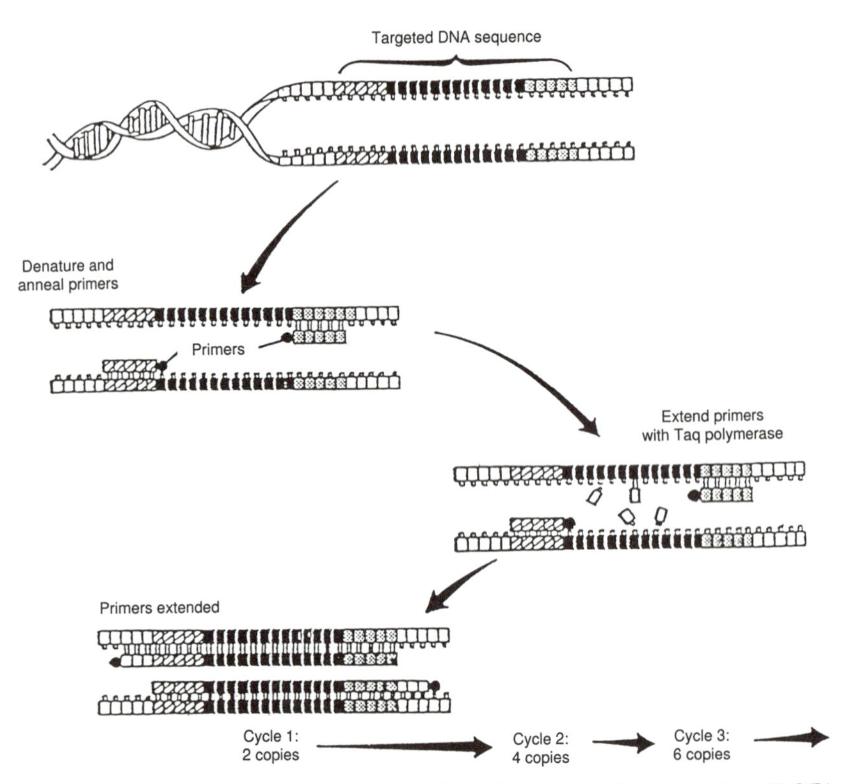

Figure 3. Amplification of DNA using the polymerase chain reaction (PCR). Each cycle consists of heat denaturation of the target molecules, annealing of an oligonucleotide primer to each target complementary DNA strand, and extension of the primers using Taq polymerase. The newly synthesized molecules are denatured to provide double the number of templates for the next cycle. Reproduced with permission from reference 3. Copyright 1992 by W. H. Freeman and Company.

the temperature is adjusted to allow the primers to anneal to the selected regions that are complementary to the sequence of the primers. Thus, one requirement for PCR is prior knowledge of the sequence of the DNA around the area to be replicated. Once the primers anneal to the target DNA, the DNA is replicated by addition of a DNA polymerase and the nucleotides dATP, dCTP, and dTTP. The denaturation process is repeated to separate the newly formed strands and a second round of replication occurs following annealing of the primers to complementary regions of both the original DNA strands as well as the newly formed strands. Repeating the cycle of denaturation, annealing, and replication leads to amplification of the section of the original DNA molecule that is bracketed by the primer set. The key to the PCR process is its use of a heat-stable DNA polymerase cloned from the thermophilic bacterium *Thermus aquaticus*.[47] This enzyme is stable at the temperatures required for heat denaturation (90–95 °C) and can withstand the variation in temperature that occurs during the PCR process.

The amplification process leads to great analytical sensitivity; it is estimated that with this technique, 100 cells provide sufficient DNA to produce a product that can be detected by hybridization to probes specific for the amplified section of the original DNA template.[8] This level of sensitivity permits genetic typing of an individual based on the cells associated with the root of a single hair.[48]

The major histocompatibility locus class II antigen DQ alpha was the first genetic system with which a PCR system was fully developed as a forensic diagnostic test. In 1987, E. T. Blake described the application of PCR to the typing of DQ alpha alleles as an approach to genetic testing of forensic evidence.[49] Reports in 1989[50–51] described further developments of PCR amplification and detection of the DQ alpha alleles in the PCR product with nonisotopically labeled allele-specific oligonucleotides. This system was tested later in 1989, and in 1990 was released by Cetus Corporation as the DQ Alpha Amplitype Kit. The kit was extensively tested in validation studies[52–55] and has become almost a standard in forensic laboratories that perform DNA-based analyses of forensic materials. In 1992, a second generation of the PCR-based forensic DNA testing system with the trade name AmpliType PM was described[56] and, in 1994, it was introduced as a kit. Following validation studies by five laboratories,[57] this kit was released by Roche Molecular Systems for general use in forensic laboratories.

Between 1988 and 1993, other PCR-based systems were described for use in analysis of forensic evidence. In 1988, A. J. Jeffreys, V. Wilson, R. Neumann, and J. Keyte described the use of PCR to detect multiallelic profiles.[58] In 1989, R. C. Allen, G. Graves, and B. Dubowle described a polyacrylamide system that allows for resolution of PCR products on polyacrylamide gels.[59] In 1990, K. Kasai, Y. Nakamura, and R. White described a method to type VNTR markers amplified by PCR;[60] this technique has been more fully developed to detect a variety of VNTR loci and given the name AMP-FLP by B. Budowle.[61–63] Another application of PCR to VNTRs is to detect STRs by both normal and automated methods.[64–66] In 1991, A. J. Jeffreys, A. MacLeod, K. Tamaki, D. L. Neil, and D. D. Monckton described a digital typing technique called *minisatellite variant repeat mapping* (MVR), which is based on the use of PCR with specially designed sets of primers that amplify a minisatellite variant.[67] In 1993, A. C. Syvanen, A. Sajantila, and M. Lukka described biallelic markers detectable by PCR and solid phase minisequencing.[68]

Types of Forensic Samples Analyzed by DNA Typing Methods

The analysis of forensic evidence by DNA typing methods requires the analyst to prepare highly purified DNA from the sample. The basic method as described by Maniatis[69] for obtaining DNA from a sample containing cellular material is to lyse the cells with a mild detergent such as sodium dodecyl sulfate (SDS), digest the proteins present with a general proteinase (usually an enzyme called proteinase K), extract non-DNA material, and concentrate the DNA by precipitation with alcohol or by ultrafiltration with membranes that retain the DNA molecules due to their large molecular mass. This technique was devised early in the development of molecular biology to obtain DNA from tissue—in particular, blood[14–18,70]—and has subsequently been adapted to analyze all types of materials collected at crime scenes, such as bloodstains, bone, hair, and semen stains.[71–75]

A specific forensic adaptation of the general method for DNA extraction was the development in 1986 of a method that could separate semen stains into the cellular contributions of the victim and the assailant. It was noted that spermatozoa could be separated from nucleated cells found in blood, secretions, and tissue because of the relative resistance of spermatozoa to lysine. Specifically, it was found that if an extract containing spermatozoa mixed with other types of cells is first lysed with mild detergents, the sperm heads remain intact and can be removed by low-speed centrifugation (5000 to 15,000 $\times$ g). The sperm then are rendered susceptible to lysis by mild detergents by the addition of the reducing agent dithiothreitol, which breaks up the thiol proteins that constitute part of the sperm head. The result is that DNA in cells in mixed stains and swabs of the vaginal cavity obtained in sexual assault cases can now potentially be separated into the victim's contribution and the assailant's contribution, providing that the sperm in the stain have intact heads.[17,71]

For RFLP analysis with VNTR markers, a requirement noted early in the development of forensic DNA testing is that the DNA isolated has to be intact, that is, undegraded, or what has been called high-molecular-weight DNA.[70] If the DNA has been extensively fragmented prior to restriction enzyme digestion, the possibility is diminished of obtaining an enzyme digest of the DNA in which the fragments released reflect the structure of the original DNA. When RFLP analysis with VNTR markers was used on forensic samples collected at crime scenes, the DNA obtained was often degraded and did not yield an RFLP pattern with either multilocus or single-locus probes. In spite of this problem, many cases still yielded DNA of sufficient quality for RFLP analysis with probes for VNTR markers. Studies conducted between 1989 and 1991 showed that degradation of the DNA does not lead to a false result in RFLP analysis. Rather, as a result of exposure to environmental effects such as bacterial contamination, sunlight, humidity, and extremes of temperature, there is loss of the signal due to degradation of the DNA but no change in the banding pattern.[73–75] Many laboratories have introduced controls to determine if bands appear because of bacterial contamination or if changes in banding patterns (often called band shifting) are caused by some effect of environment on the DNA.

Because physical forensic evidence often consists of multiple small stains or only a few hairs, it would facilitate the analysis if smaller quantities could be used. This matter was addressed by the development of PCR methodology that yields sufficient DNA for PCR amplification from as few as 100 cells.[8] Thus, a single live hair that still has roots attached will often yield a genetic typing result.[48] Degradation is less important also for PCR methods because only small sections of DNA (100 to 500 base pairs) are amplified; with such a small target (the entire human genome contains about 3×10^9 base pairs), there is less chance that degradation will occur at a site critical to PCR amplification. The PCR method, however, will only work when there is an intact stretch of DNA between the two sites where the primers anneal.

With the introduction of PCR for analysis of forensic evidence, attention has focused on development of methods for isolation of DNA from cells found in trace biological samples. The most notable advance has been the development by P. S. Walsh et al. of a method to absorb DNA from a sample onto an ion exchange resin called Chelex, followed by elution with boiling.[76] This method has been further refined by M. N. Hochmeister, B. Budowle, and their colleagues to allow for extraction of DNA from epithelial cells that may be found on cigarette butts, envelopes, or postage stamps.[77]

Studies have been made of the effects on PCR of the environment and of chemicals that may be found in association with forensic evidence. No incorrect typings have been noted, but there have been losses of signal from either loss of DNA or inhibition of the Taq DNA polymerase enzyme.[54] In contrast to RFLP analysis, it has been noted that with PCR analysis casual contact of samples with each other can lead to transfer of sufficient DNA to allow for detection, which raises the possibility of an incorrect type.[78] More importantly, because the small sections of amplified DNA are preferentially amplified in the PCR reaction, it is important to segregate all samples of forensic evidence from any potential source of amplified human DNA. Thus, containment measures have been developed and recommended for use in the laboratory to ensure that there is no cross contamination between samples either before or after amplification.[79] The issue of contamination of samples collected at crime scenes is just beginning to be addressed with crime scene investigators.[80]

In spite of these technical problems, crime laboratories that perform PCR DNA analysis have successfully developed protocols for isolating and typing DNA from a plethora of types of samples. A partial list includes stains found on almost any surface except soil[76,81]; tissues on stained slides and contained in paraffin blocks prepared for pathological analysis;[82,83] formalin-fixed tissue;[70] bones and teeth from exhumed bodies;[84–85] bone, teeth, and tissue from burned bodies;[85] blood and semen samples over 40 years old;[86] saliva containing epithelial cells obtained from envelopes;[87] bite marks and cigarette butts;[77] and single live hairs.[48] It appears that the types of samples that can be analyzed by PCR methodology is limited only by the technical requirements for extraction of the DNA, provided there is sufficient quantity and quality of DNA to be isolated.

Genetic Markers Detected by DNA Typing Methods

Table I lists PCR assay formats and types of genetic systems that have been or are being evaluated for use with DNA typing methodology for analysis of forensic evidence. Some of these systems, most notably the DQ Alpha

Table I. PCR Assays Used in DNA Forensic Analysis.

Assay Format	*Detection Systems*	*Genetic Systems Detected*	*Test*
Reverse dot blot	Hybridization to allele-specific oligonucleotides	Single-copy genes	DQ Alpha Amplitype kit AmpliType PM kit
Direct dot blot	Hybridization to sequence-specific oligonucleotides	Single-copy genes	MHC class II antigen typing
Electrophoretic analysis	Silver staining of polyacrylamide gels	VNTR markers STR markers	AMP–FLP
Southern blotting	Hybridization to VNTR probe	VNTR markers	MVR
Solid phase minisequencing	Hybridization to biallelic probes	STR markers	Biallelic analysis

AmpliType and AmpliType PM test kits, are based on PCR amplifications followed by detection of single-copy genes that code for amino acid sequences of known proteins. The genetic systems analyzed in these kits have been described previously by serological and eletrophoretic methods. The PCR systems based on VNTR markers require characterization in well-defined populations before they can be used for analysis of forensic evidence.

The VNTR markers used most frequently in RFLP analyses of forensic evidence are found in the noncoding regions or introns of DNA.[26] Because they code for no known function, no protein product has been associated with these markers. These regions exhibit genetic polymorphism as described previously. Jeffreys has suggested that one function of the tandem repeated DNA sequences is as a site or "hot spot" for crossover events that occur during meiosis.[88] VNTR markers have been found to be dispersed throughout the human genome, but in forensic testing the focus has been on about 10 genomes, all found on different chromosomes. Table II lists the VNTR markers most commonly used in forensic analysis and references to the articles in which they were first described.

Genetic markers detected by PCR in forensic testing are both single-copy genes of known function and also tandemly repeated short sequences of DNA. Thus, the DQ Alpha AmpliType test detects six of the ten known

Table II. Single-Locus VNTR Markers Used in RFLP Analysis for Forensic DNA Testing

Marker	*Reference*
D2S44	*117*
D14S13	*118*
D10S28	*119*
D17S79	*120*
D1S7	*121*
D4S139	*122*
3'HVR	*123*
D18S27	*124*
D55110	*125*

alleles that constitute the DQ alpha locus that codes for the alpha chain of the DQ alpha heterodimer protein, one of the class II antigens that constitute the major histocompatibility locus on chromosomes.[89] Similarly, the AmpliType PM test for alleles for glycophorin A (GYPA), low-density lipoprotein receptor (LDLR), group-specific globulin (GC), and hemoglobin G-gamma globulin (HBGG).[90–93] There are also PCR tests for short-tandem repeats that are part of the apolipoprotein B (APO-B) and (HUM THO1) loci[94,95] as well as trimeric and tetrameric repeats.[96] Short-tandem repeats that are more similar to the VNTR include D1 S80 and D7 S8.[60,63,97] Recently, a PCR method has been described that can genotype ABO blood group antigens, which increases the number of ABO phenotypes detected from three to five.[98]

One DNA typing technique that represents an important advance in forensic testing is the ability to determine the sex of donors of biological materials that are part of forensic evidence. Early in the development of RFLP testing, probes were developed by Lifecodes that allow detection of X and Y chromosome–specific DNA; similar probes are in wide use in many laboratories, particularly in Canada. Recently, a major effort has been directed toward developing PCR methods for determining the sex of the donor of extracted DNA. Markers that hold great promise include ameliogenin and a zinc finger protein.[99,100] One DNA technology that has yet to be explored, but that could prove very useful in analysis of forensic evidence, is the fluorescent in situ hybridization (FISH) methodology used widely in cytogenetic laboratories to determine the sex of cells.[101]

Scientific and Legal Acceptance of the Results of DNA Typing of Forensic Evidence

Issues concerning the scientific and legal acceptance of DNA typing of forensic evidence were publicly discussed at a conference held in late 1988 at Cold Spring Harbor and published in 1989 in Banbury Report 32.[102] During 1989, the first significant scientific and legal challenge to the use of DNA analysis of forensic evidence came about in *New York v. Castro;* in that case, RFLP analysis had been performed with multiband and single-locus probes.[103] The court found general scientific acceptance of RFLP analysis but in this particular case excluded test results of the DNA match because of problems with its application. This case was particularly important; from it came a three-pronged approach to determining the admissibility of the results of DNA analysis: Is there a theory, generally accepted in the scientific community, that supports the conclusion that DNA forensic testing can produce reliable results? Are there currently techniques or experiments that produce reliable results in DNA identification and that are generally accepted in the scientific community? Did the laboratory perform the test using accepted scientific techniques to analyze the forensic samples in this particular case?

In this context, the scientific questions and legal challenges to DNA forensic analyses based on either RFLP or PCR test results since 1989 have generally focused on the following areas.

Methodology Used

Forensic DNA testing now uses both RFLP and PCR methods; the theory behind both methods has been proven in many laboratories. Both are key technologies used in the analysis of DNA for genetic markers.[11,46]

Specific Applications of RFLP and PCR Methods for Use with Forensic Samples

Reagents qualified for forensic RFLP analysis can be obtained from several suppliers,[104] standard preparations of control DNA preparations are now available from the National Institutes of Standards of Technology (NIST) in Gaithersburg, Maryland; the protocols for RFLP analysis have been reviewed both in publications and in the course of numerous discovery motions; and the population databases have been made widely available for analysis by other scientists. For PCR, a kit for analysis of forensic samples that detects the DQ alpha alleles has been developed and is available commercially (the DQ-Alpha AmpliType kit from Perkin Elmer of Norwalk, Connecticut). A second-generation kit (AmpliType PM from Roche Molecular Systems) was introduced in 1994. Both meet the requirements of government regulatory agencies for kits that are to be used for medical diagnoses.

Accuracy and Reproducibility of the Data

With RFLP analysis, questions were raised about the difficulties of precisely measuring DNA bands that might differ by less than 50 base pairs and about conditions that could lead to shifting of bands. With PCR, questions were asked about the mistake rate of the Taq DNA polymerase in incorporating the correct base and about hybridization conditions that allow for distinguishing DQ alpha alleles that differ by a single base pair. Most such questions have been answered by direct experimentation and by documentation of the testing done by laboratories conforming to standard operating protocols that allow for validation of such measurements.

Determination of Frequencies of Markers or Alleles

For most VNTR markers, the total number of different markers at each locus used has yet to be precisely defined. The frequencies of these markers (that is, the molecular weights of bands observed) appear, however, to follow a continuous distribution that for some markers seems to be bimodal.[25,28,43] The solution to this problem is to make bins into which a marker is placed and then calculate the frequencies based on the number of observations that occur in a bin. The FBI chose to use fixed bins, the boundaries of which are determined with molecular-size standard markers and for which the boundaries are greater than the determined measurement imprecision of the analytical system. The imprecision was determined empirically to be ±2.5%. Bins are combined when the number of markers falling in a bin is less than 5.[105]

Lifecodes chose to use floating bins, where the frequency of the marker is calculated based on the number of events that occur within 3 standard deviations of the measurement, which in the Lifecodes method is ±1.8% of the observed molecular weight.[25] Cellmark developed a system for single-locus probes similar to that of Lifecodes, except that measurement standards were set at 1 mm for comparing bands on the same gel and 2 mm for comparing bands on different gels. With the PCR method for the DQ alpha alleles, the total number of DQ alpha alleles is 15, the sequence of the DNA for each allele is known, and standard statistical methods for evaluating phenotype frequencies and calculating genotype frequencies are used.[52] The results of the initial population data on the DQ alpha genotype frequencies have been substantiated by studies in several laboratories.[54,106,107] Roche Molecular Systems has developed similar data for the AmpliType PM system.

Genetic Properties of the Systems

Many of the VNTR markers initially appeared to exhibit deviations from Hardy–Weinberg equilibrium and also seemed to show linkage disequilibrium.[30,31,108–112] Another criticism was that because the ethnic groups used in the initial studies were mixtures of diverse subgroups, a Whalund effect would occur, namely, there would be an excess of single (e.g., apparent) homozygotes.[31,108] The response to these criticisms was twofold. First, the size of the database was increased and data were collected from identical ethnic groups in different geographic locations.[28] Also, an RFLP compendium was created from DNA samples collected around the world.[113] Second, the cause of the apparent excess of homozygosity in the Lifecodes database was analyzed and found to be explained by the coalescence of two bands into one band.[35] In the case of the FBI database, null markers apparently produce most of the single-band phenotypes.[38]

The conclusion: the VNTR systems used in forensic analysis do segregate independently. Modeling of the data has indicated that allele distributions are not that different for ethnic groups and it has been suggested that if there are subpopulations, it would make very little difference in the observed frequencies.[33,37] This criticism of the PCR DQ Alpha method has not yet been raised. Also, H. Erlich has discussed why the ceiling principle does not apply to the class II antigens.[114]

Establishment of Standards and Accreditation for Forensic Laboratories

This aspect of DNA forensic testing has been summarized and addressed in *Genetic Witness: Forensic Uses of DNA Tests,*[12] a report from the Office of Technology Assessment, and also in a report from the National Research Council, *DNA Technology in Forensic Science.*[39] Furthermore, legislation has been proposed, the DNA Proficiency Act of 1991.[115] In the meantime, the American Association of Blood Banks has established rules for accrediting laboratories to do DNA analysis for paternity testing, the American Society of Crime Laboratory Directors accredits forensic laboratories for DNA testing, and an informal group, the Technical Working Group on DNA Analysis, has promulgated guidelines on laboratories on both RFLP and PCR analysis. In 1994, the American Society for Histocompatibility also began accrediting laboratories for PCR testing.

The legal acceptance of DNA testing in the United States continues to expand. M. D. Stolorow and G. W. Clark[116] as well as J. T. Sylvester[103] have reviewed this area.[88] In April 1992, it was reported that DNA analysis had been conducted in over 14,700 criminal cases and admitted in over 612 criminal trials. The majority of these were DQ Alpha AmpliType results. Overall, DNA results were not admitted in seven reported and four unreported cases. No exact statistics are available, but a survey of the literature indicates that DNA testing is widely accepted in western Europe as well as in Israel, India, and Japan.

Conclusion

This chapter focused on the scientific development of DNA testing for forensics and the response of the forensic scientist to the scientific and legal communities' questions about this new approach to the analysis of biological evidence collected at crime scenes. It is widely assumed that in the next few years DNA testing will replace most of the conventional tests now used to analyze blood, semen, and other body tissues. Yet, even as the methods of DNA forensic testing continue to change, forensic scientists will search for ways to further increase and refine the tests' analytical sensitivity and specificity. All these activities seek the goal of completing the transformation of what was recently a research technique into a routine procedure that addresses all concerns for the legal and ethical issues that are involved.

Acknowledgments

I would like to acknowledge the excellent word processing assistance of Helen Hourihan and her staff at the Center for Blood Research, and the support of Janice Williamson, Susan Mitchell, Rafaat Houranieh, Debra Mozill, and Richard L. Korn of the CBR Laboratories.

References

1. Avery, O. T.; MacCleod, C. H.; McCarty, M. *J. Exp. Med.* **1944,** *79,* 137–158.
2. Sensabaugh, G. F. In *Forensic Science Handbook;* Saferstein, R., Ed.; Prentice-Hall: Englewood Cliffs, 1982; Chapter 8, pp 338–415.
3. Kirby, L. T. *DNA Finger Printing;* Stockton: New York, 1990; pp 7–33.
4. Alberts, B.; Bray, D.; Lewis, J.; Raff, M.; Roberts, K.; Watson, J. D. *Molecular Biology of the Cell,* 2nd ed.; Garland: New York, 1989; Chapter 10, pp 551–556.
5. Paabo, S. *Proc. Natl. Acad. Sci. U.S.A.* **1989,** *86,* 1939–1943.
6. Lawlor, D. A.; Dickel, C. D.; Hauswith, W. W.; Parhams, P. *Nature (London)* **1991,** *349,* 785–787.
7. DeSalle, R.; Batesy, J.; Wheeler, W.; Grimaldi, D. *Science (Washington, DC)* **1992,** *257,* 1933–1936.
8. Reynolds, R.; Sensabaugh, G.; Blake, E. *Anal. Chem.* **1991,** *63,* 2–15.
9. Southern, E. M. *J. Mol. Biol.* **1975,** *98,* 503–517.
10. Elder, J. K.; Southern, E. M. *Anal. Biochem.* **1983,** *128,* 227–231.
11. Botstein, D.; White, R. L.; Skolnick, M.; Davis, R. W. *Am. J. Hum. Genet.* **1980,** *32,* 314–331.
12. *Genetic Witness: Forensic Uses of DNA Tests;* U.S. Congress, Office of Technology Assessment. U.S. Government Printing Office: Washington, DC, 1990; OTA-BA 438.
13. Wynan, A. R.; White, R. L. *Proc. Natl. Acad. Sci. U.S.A.* **1980,** *77,* 6754–6758.
14. Fowler, J. C. S.; Burgoyne, L. A.; Scott, A. C.; Harding, H. W. J. *J. Forensic Sci.* **1988,** *33,* 1111–1126.
15. Jeffreys, A. J.; Wilson, V.; Thein, S. L. *Nature (London)* **1985,** *316,* 76–79.
16. Jeffreys, A. J.; Brookfield, J. F. Y.; Semenoff, R. *Nature (London)* **1985,** *317,* 818–819.
17. Gill, P.; Jeffreys, A. J.; Werrett, D. J. *Nature (London)* **1985,** *318,* 577–579.
18. Hagelberg, E.; Gray, F. C.; Jeffreys, A. J. *Nature (London)* **1991,** *352,* 427–429.
19. Jeffreys, A. J.; Turner, M.; Debenham, P. *Am. J. Hum. Genet.* **1991,** *48,* 824–840.
20. Jeffreys, A. J.; Wilson, V.; Thein, S. L. *Nature (London)* **1985,** *314,* 67–73.
21. Budowle, B. *Crime and Laboratory Digest* **1988,** *15,* 97–98.
22. Nakamura, Y.; Leppert, M.; O'Connell, P.; Wolf, R.; Holm, T.; Culver, M.; Fujimoto, E.; Hoff, M.; Kumlin, E.; White, R. *Science (Washington, D.C.)* **1987,** *235,* 1616–1622.
23. Fowler, S. J.; Gill, P.; Werrett, D. J.; Higgs, D. R. *Hum. Genet.* **1988,** *79,* 142–146.
24. Odelberg, S. J.; Plaetke, R.; Eldridge, J. R.; Ballard, L.; O'Connell, P. O.; Nakamura, Y.; Leppert, M.; Lalouel, J.-M.; White, R. *Genomics* **1989,** *5,* 915–924.
25. Balazs, I.; Baird, M.; Clyne, M.; Meade, E. *Am. J. Hum. Genet.* **1989,** *44,* 182–190.
26. Cawood, A. H. *Clin. Chem.* **1989,** *35,* 1832–1837.
27. Odelberg, S. J.; White, R. In *DNA and Other Polymorphisms in Forensic Science.* Lee, H. C.; Gaesslein, R. E., Eds.; Adv. in Forensic Sciences.; 1990; Vol. 3, Chapter 2.
28. Budowle, B.; Monson, K. L.; Anoe, K. S.; Baechtel, S.; Bergman, D. C.; et al. *Crime Laboratory Digest* **1991,** *18,* 9–28.
29. Budowle, B.; Giusti, A. M.; Waye, J. S.; Baechetel, F. S.; Fourney, R. M.; Adams, D. E.; Preseley, L. A.; Monson, K. L. *Am. J. Hum. Genet.* **1991,** *48,* 841–855.
30. Charaborty, R.; de Andrade, M.; Daiger, S. P.; Budowle, B. *Am. J. Human Genet.* **1992,** *56,* 45–47.
31. Nei, M. *Molecular Evolutionary Genetics;* Columbia University Press: New York, 1987; Chapter 8.

32. Chakraborty, R.; Jin, L. *Human Genet.* **1992,** *88,* 267–272.
33. Chakraborty, R.; Kidd, K. K. *Science (Washington, D.C.)* **1991,** *254,* 1735–1739.
34. Balazs, I.; Baird, M. L.; McElfresh, K.; Udey, J. *Adv. in Forensic Hemogenetics* **1990,** *3,* 71–74.
35. Devlin, B.; Risch, N.; Roeder, K. *Science (Washington, D.C.)* **1990,** *249,* 1416–1420.
36. Devlin, B.; Risch, N.; Roeder, K. *Am. J. Hum. Genet.* **1991,** *43,* 662–676.
37. Devlin, B.; Risch, N. *Am. J. Hum. Gent.* **1992,** *51,* 534–548.
38. Devlin, B.; Risch, N.; Roeder, K. *Science* **1993,** 7480–749.
39. National Research Council. *DNA Technology in Forensic Science;* National Academy of Sciences: Washington, DC, 1992; Chapter 3.
40. National Research Council. *The Evaluation of Forensic DNA Evidence;* National Academy of Sciences: Washington, DC, 1996; Executive Summary: Overview.
41. Budowle, B.; Lander, E. S. *Nature* **1994,** *371,* 735–738.
42. Weir, B. *Am. J. Hum. Genet.* **1996,** *59,* 497–500.
43. Deka, R. R.; Chakraborty, R.; Ferrell, R. E. *Genomics* **1991,** *11,* 83–92.
44. Balazs, I.; Nemweiler, J.; Gunn, P.; Kidd, J.; Kidd, K. K.; Kuhl, J.; Mingjun, L. *Genetics* **1992,** *131,* 191–198.
45. Edwards, A.; Civitello, A.; Hammond, H. A.; Caskey, C. T. *Am. J. Hum. Genet.* **1991,** *49,* 746–756.
46. Saiki, R. K.; Bugawan, T. L.; Horn, G. T.; Mullis, K. B.; Erlich, H. A. *Nature (London)* **1986,** *324,* 163–166.
47. Saiki, R. K.; Gelfanel, D. H.; Stoffel, S.; Scharf, S. J.; Higuchi, R.; Horn, G. T.; Mullis, K. B.; Erlich, H. A. *Science* **1988,** *239,* 487–491.
48. Higuchi, R.; von Beroldingen, C. H.; Sensabaugh, G.; Erlich, H. A. *Nature (London)* **1988,** *332,* 543–546.
49. Blake, E. T., personal communications.
50. Higuchi, R.; Blake, E. T. In *DNA Technology and Forensic Science;* Ballantyne, J.; Sensabaugh, G.; Witkowski, J., Eds.; Banbury Report 32; Cold Spring Harbor Laboratory: New York, 1988; pp 265–281.
51. Saiki, R. K.; Walsh, P. S.; Levenson, C. H.; Erlich, H. A. *Proc. Natl. Acad. Sci. U. S. A.* **1989,** *86,* 6230–6234.
52. Helmuth, R.; Fildes, N.; Blake, E.; Luce, M. C.; Chimera, J.; Madej, R.; Gorodezky, C.; Stone king, M.; Schmill, N.; Klitz, W.; Higudi, R.; Erlich, H. A. *Am. J. Hum. Gent.* **1990,** *47,* 515–523.
53. Walsh, P. S.; Fildes, N.; Louie, A. S.; Higudi, R. *J. Forensic Sci.* **1991,** *36,* 1551–1556.
54. Comey, C. T.; Budowle, B. *J. Forensic. Sci.* **1991,** *36,* 1633–1648.
55. Blake, E.; Mihalovich, J.; Higudi, R.; Walsh, S. P.; Erlich, H. A. *J. Forensic Sci.* **1992,** *37,* 700–726.
56. Reynolds, R. *Crime Laboratory Digest* **1991,** *18,* 132–133.
57. Fildes, N., Reynolds, R. *J. Forensic Sci.* **1995,** *40,* 279–288.
58. Jeffreys, A. J.; Wilson, V.; Newman, R.; Keyte, J. *Nucl. Acids Res.* **1988,** *16,* 10953–10974.
59. Allen, R. C.; Graves, G.; Budowle, B. *BioTechniques* **1989,** *7,* 736–744.
60. Kasai, K.; Nakamura, Y.; White, R. *J. Forensic Sci.* **1990,** *35,* 1196–1200.
61. Comey, C. T.; Alevy, M. C.; Parsons, G. L.; Wilson, M.; Budowle, B. In *Proc. Sec. and Intl. Symp. Human Identification;* Promega Corp.: Madison, WI, 1991; pp 53–62.
62. Budowle, B. *Crime Laboratory Digest* **1991,** *18,* 134–134.

63. Budowle, B.; Chakraborty, R.; Giusti, A. M.; Eisenberg, A. J.; Allen, R. C. *Am. J. Hum. Genet.* **1991,** *48,* 137–144.
64. Edwards, A.; Civatello, A.; Hammond, H. A.; Caskey, C. T. *Am. J. Hum. Genet.* **1991,** *49,* 746–756.
65. Caskey, C. T.; Pizzuti, A.; Fu, Y.-H.; Fenwick, R. G., Jr.; Nelson, D. L. *Science* **1992,** *256,* 784–788.
66. Hammond, H. A.; Caskey, C. T. *Proc. Third Intl. Symp. Human Identification;* Promega: Madison, WI, 1992; pp 163–175.
67. Jeffreys, A. J.; MacLeod, A.; Tamaki, K.; Neil, D. L.; Monckton, D. D. *Nature (London)* **1991,** *354,* 204–209.
68. Syvanen, A.-C.; Sajantila, A.; Lukka, M. *Am. J. Hum. Genet.* **1993,** *32,* 46–59.
69. Maniatis, T.; Fritsch, E. F.; Sambrook, J. *Molecular Cloning: A Laboratory Manual;* Cold Spring Harbor Laboratory: New York, 1982.
70. Kanter, F.; Baird, M.; Shaler, R.; Balazs, I. *J. Forensic Sci.* **1986,** *31,* 402–408.
71. Giusti, A.; Baird, M.; Pasguale, S.; Balazs, I.; Glassberg, J. *J. Forensic Sci.* **1986,** *31,* 409–417.
72. Hochmeister, M. N.; Budowle, B.; Borer, U. V.; Eggmann, U.; Comey, C. T.; Dirnhofer, R. *J. Forensic Sci.* **1991,** *36,* 1649–1661.
73. McNally, L.; Shaler, R. C.; Baird, M.; Balazs, I.; DeForest, P.; Kobilinsky, L. *J. Forensic Sci.* **1989,** *34,* 10559–1069.
74. McNally, L.; Shaler, R. C.; Baird, M.; Balazs, I.; Kobilinsky, L.; DeForest, P. *J. Forensic Sci.* **1989,** *34,* 1070–1077.
75. Adams, D. E.; Presley, L. A.; Baumstark, A. L.; Henstey, K. W.; Hill, A. L.; Aroe, K. S.; Campbell, P. A.; McLaughlin, C. M.; Budowle, B.; Giusti, A. M.; Smerick, J. B.; Baechtel, F. S. *J. Forensic Sci.* **1991,** *36,* 1284–1296.
76. Walsh, P. S.; Metzger, D. A.; Higuchi, R. *BioTechniques* **1991,** *10,* 506–513.
77. Hochmeister, M. N.; Budowle, B.; Jung, J.; Borer, U. V.; Comey, C. T.; Dirnhofer, R. *Int. J. Leg. Med.* **1991,** *104,* 229–233.
78. *AmpliType™ User Guide Version 2*. Cetus Corp.: Emeryville, CA, 1990.
79. Orrego, C. In: *PCR Protocols, A Guide to Methods and Application;* Innis, M. A.; Gelfand, D.; Sninsky, J. J.; White, T. J., Eds.; Academic: New York, 1990; Chapter 54.
80. Lee, H. C., Gaensslon, R. E., Bigbee, P. D., Kearney, I. J. *J. S. Dept. of Justice, Federal Bureau of Investigation*, 1993.
81. Jung, J. M.; Comey, C. T.; Baer, D.; Budowle, B. *Int. J. Leg. Med.* **1991,** *104,* 145–148.
82. Impraim, C.; Saiki, R. K.; Erlich, H. A.; Teplitz, R. L. *Biochem. Biophys. Res. Comm.* **1987,** *142,* 710–716.
83. Shibata, D.; Namiki, T.; Higuchi, R. *Amer. J. Surg. Pathol.* **1990,** *73,* 1–3.
84. Hochmeister, M. N.; Budowle, B.; Borer, U. V.; Eggmann, U.; Comey, C. T.; Diruhofer, R. *J. Forensic Sci.* **1991,** *36,* 1649–1661.
85. Lee, H. C.; Pagliaro, E. M.; Gaensslen, R. E. Berka; K. M.; Keith, T. P.; Keith, G. N.; Garner, D. D. *J. Forensic Sci. Soc.* **1991,** *31,* 209–212.
86. Bing, D. H.; Duseman, B. 1993. Unpublished casework.
87. *F.B.I. DQα Alpha Manual,* Revised 6/92. FBI Academy: Quantico, VA
88. Royle, N. J.; Clarkson, R. E.; Wong, Z.; Jeffreys, A. J. *Genomics* **1988,** *3,* 352–360.
89. Imanishi, T., Akaza, T., Kimura, A., Tokunga, K., Gojobori, T. In *HLA* **1991,** *Proceedings of Eleventh International Histocompatability Workshop and Conference;* Tsuji, K.; Aizawa, M.; Susazuki, T., Eds.; Oxford University: New York, 1992.

90. Yamamoto, T.; Davis, C. G.; Brown, M. S.; Schneider, W. J.; Casey, M. L.; Goldstein, J. L.; Russell, D. W. *Cell* **1984,** *39,* 27–28.
91. Siebert, P. D.; Rukada, M. *Proc. Nat. Acad. Sci. U. S. A.* **1987,** *84,* 6735–6739.
92. Slightom, J. L.; Blechl, A. E.; Smithies, O. *Cell* **1980,** *21,* 627–638.
93. Yang, F.; Brune, J. L.; Naylor, S. L.; Apples, R. L.; Naberhaus, K. H. *Proc. Natl. Acad. Sci. U. S. A.* **1985,** *82,* 7994–7998.
94. Boerwinkle, E.; Xiong, W.; Fourest, E.; Chan, L. *Proc. Nat. Acad. Sci. U.S.A.* **1989,** *86,* 212–216.
95. Puers, C.; Hammond, H. A.; Jin, L.; Caskey, C. T.; Shumm, J. W. *Am. J. Hum. Genet.* **1993,** *53,* 953–958.
96. Edwards, A.; Hammond, H. A.; Jin, L.; Caskey, C. T.; Chakraborty, R. *Genomics* **1992,** *12,* 241–253.
97. Horn, G. T.; Richards, B.; Merrill, J. J.; Klinger, K. W. *Cell* **1980,** *21,* 627–638.
98. Johnson, P. H.; Hopkinson, D. A. *Hum. Mol. Genetics* **1992,** *1,* 341–344.
99. Mannucci, A.; DeStafano, F.; Casarino, B. S.; Canale, M. *Amer. Acad. For. Sci.* **1994,** *Abstract B65.*
100. Varlano, J.; Reynolds, R. *Amer. Acad. For. Sci.* **1994,** *Abstract B75.*
101. Pinkel, D., Straume, T., Gray, J.W. *Proc. Nat. Acad. Sci. U.S.A.* **1986,** *83,* 2934–2938.
102. *DNA Technology and Forensic Science;* Ballantyne, J.; Sensabaugh, G.; Witowski, J., Eds.; Banbury Report 32; Cold Spring Harbor Laboratory, NY, 1989.
103. Sylvester, J. T. *Third Intl. Symp. on Hum. Identification;* Promega Corp.: Madison, WI, 1992; pp 61–83.
104. Analytical Genetic Testing Center, Denver, CO; BRL, a Division of Gibco Life Technologies, Gaithersburg, MD; Cellmark Corp., Germantown, MD; Genelex Corp., Seattle, WA; Lifecodes Corp., Stanford, CT; Perkin-Elmer Corp. Applied BioSystems, Norwalk, CT; Promega Corp., Madison, WI.
105. Budowle, B.; Giusti, A. M.; Waye, J. S.; Baechtel, F. S.; Fourney, R. M.; Adams, D. E.; Presley, L. A.; et al. *Am. J. Hum. Genet.* **1991,** *48,* 841–855.
106. Ronningen, K. S.; Spurkland, A.; Markussen, G.; Iwe, T.; Vartelal, F.; Thorsby, E. *Hum. Immunol.* **1990,** *29,* 275–281.
107. Sajantila, A.; Strom, M.; Budowle, B; Tienari, P. J.; Ehnholm, C.; Peltonen, L. *Int. J. Leg. Med.* **1991,** *104,* 181–184.
108. Cohen, J. E. *Am. J. Hum. Genet.* **1990,** *46,* 358–368.
109. Lander, E. *Amer. J. Hum. Genet.* **1991,** *48,* 819–823.
110. Caskey, C. T. *Amer. J. Human Genet.* **1991,** *49,* 893–895.
111. Cohen, J. E.; Lynch, M.; Taylor. *Science* **1991,** *243,* 1037–1038.
112. This topic has been addressed in a series of letters published in the *Am. J. Hum. Genet.* and in a section or technical comment published in *Science.*
113. Monson, K. L.; Budowle, B. *Amer. Acad. For. Sci.* **1993,** *Abstract B101,* 84.
114. Erlich, H. A. *Amer. Acad. For. Sci.* 1993.
115. Hicks, J. W. *Crime Lab. Digest.* **1991,** *18,* 3–8.
116. Stolorow, M. D.; Clark, G. W. *Prosecutor* **1992,** 13–28.
117. Nakamura, Y.; Gillilan, S.; O'Connell, P.; Leppert, M.; Lathrop, G. M.; Lalovel, J. M.; White, R. *Nucleic Acids Res.* **1988,** *16,* 381.
118. Nakamura, Y.; Culver, M.; Gill, J.; O'Connell, P.; Leppert, M.; Lathrop, G. M.; Lalovel, J.-M.; White, R. *Nulceic Acids Res.* **1988,** *16,* 381.
119. Bragg, T.; Nakamura, Y.; Jones, C.; White, R. *Nucleic Acids Res.* **1988,** *16,* 11395.

120. Nakamura, Y.; Lathrop, M.; O'Connell, P.; Leppert, M.; Barker, D.; Wright, E.; Skolnick, M.; Knodoleon, S.; Lalovel J.-M. *Genomics* **1988,** *2,* 302–309.
121. Wong, Z.; Wilson, V.; Patel, I.; Povey, S.; Jeffreys, A. J. *Ann. Hum. Genet.* **1987,** *51,* 269–288.
122. Milner, E. C. B.; Lotshow, C. L.; Willems van Dijk, K.; Charmley, P.; Concannon, P.; Schroeder, Jr., H. W. *Nucleic Acids Res.* **1989,** *17,* 4002.
123. Fowler, S. J.; Gill, P.; Werrett, D. J.; Jiggs, D. R. *Human Genet.* **1988,** *79,* 142–146.
124. Ip, N. J.; van de Stadt, I.; Loewy, G. Z.; Leary, S.; Grzeschik, K.–H.; Balazs, I. *Nucleic Acids Res.* **1989,** *17,* 8404.
125. Armour, J. A. *Nucleic Acids Res.* **1990,** *8,* 501–502.

4

Forensic Dust Analysis

Nicholas Petraco

A century ago, Hans Gross, a magistrate, speculated in his writing that dust is a representation of our environment in miniature. Gross further proposed that by recognizing the constituents composing a particular dust sample one could estimate the surroundings from which the dust originated, and that this information could be used to help solve crimes.[1] About the same time the importance of trace evidential material in solving crimes was being popularized by Sir Arthur Conan Doyle through his fictional character, Sherlock Holmes, who solved many mysteries by reconstructing the events of the crime from dust traces left at or taken away from the crime scene. Gross's and Conan Doyle's writings are believed to have inspired many a European and American scientific detective to look for valuable dust clues while investigating crimes.

In 1910 Edmond Locard established a police laboratory in Lyons, France, and soon his ability to solve crimes by analyzing the dust found on a suspect became known throughout the world. Locard's successful implementation of scientific methodology in criminal investigations and his belief that dust analysis could link every criminal to his or her crime are probably responsible for the use of trace evidence in contemporary times. Moreover, his work served to spur the development of forensic laboratories throughout the world.[2]

One such laboratory was developed in Sweden in the 1920s by Harry Söderman, a student and protégé of Locard. During his career Söderman advocated the use of dust analysis in scientific criminal investigations. He went on to become the chief director of Sweden's National Institute of Technical Police and was very influential in the International Association of Chiefs of Police.

In Zurich, Switzerland, during the 1950s Max Frei-Sulzer was instrumental in establishing a police laboratory that quickly gained international recognition for the examination of microtraces of such evidential materials as pollen, fibers, hair, dust, and soil. Some of Frei-Sulzer's cases involving the analysis of microscopic traces are legendary in the annals of scientific criminal investigation.

Between 1920 and 1950, forensic laboratories were developed in England, Germany, and Austria as well as the United States. By the early 1930s a national crime laboratory was maintained by the Federal Bureau of Investigation (FBI) in Washington, D.C. Scientific crime detection laboratories were also developed in large urban areas by progressive police officials. The Los Angeles police chief August Vollmer is credited with establishing the first forensic laboratory in the United States. In New York City, a well-equipped crime laboratory was in operation by the early 1930s. From their very inception these laboratories, influenced by Locard, advocated the analysis of dust and its components in routine criminal investigation.

With the development of forensic laboratories throughout the world came the establishment of schools of criminology and criminalistics. A few of the more noted forensic science programs were established at the University of Lyons by Locard, the University of California at Berkeley by Vollmer, and the City University of New York, John Jay Center of Criminal Justice, by Alexander Josephs. Due to the work and teachings of these early pioneers and many other forensic scientists past and present, the study of microscopy, microtechniques, and the identification of trace evidential materials such as dust became a vital component of these forensic science programs. In turn, they served as the impetus for succeeding generations of forensic investigators.

During my own tenure at the New York City police laboratory, much of my work involved the study of forensic dust specimens. The valuable data gained have allowed me to associate the people, places, and things involved in crimes; to deduce the occupation(s) of the principal(s) involved in crimes; and to reconstruct the crime scene and sometimes the crime itself. The goal of this chapter is to introduce the reader to the analysis of the materials commonly found in forensic dust specimens (see Table I for a list of such materials), and then show how the information has been used in actual cases to help solve crimes.

The procedures employed in the collection and preservation of the various items of physical evidence to be examined are important to the successful analysis of forensic dust specimens. The associations made possible by traces are based primarily on the mutual exchange principle attributed to Locard by Nickolls.[3] Locard's original work sets forth the basic elements of this hypothesis, which postulates that "whenever two people, places, and/or things interact there is always a mutual cross transfer of trace materials from one to the other".[4]

The trace materials transferred during these contacts make the stated associations and deductions possible. Accidental contact between the items of physical evidence to be processed for trace evidential materials must be guarded against to prevent contamination. Each item of physical evidence must be kept separate. This is easily accomplished by wrapping each item individually in paper, or by placing each in a separate paper container.

Upon receipt of physical evidence at the laboratory, the dust traces must be collected from each item. A systematic approach is vital when retrieving dust traces. Each item of physical evidence should be processed separately, using the following procedure.

The item to be examined should be removed from its container and laid out on a clean piece of paper placed on a well-lighted examination

Table I. Some of the Substances Found in Dust Specimens.

Form	*Materials*
Fiber	Human hair and animal hair, synthetic fibers, mineral and glass fibers, wood fibers, duck and chicken feathers
Particle	Fragments of minerals, glass chips, paint flakes and smears, wood dust, wood splinters and chips, bark twigs, leaf fragments, pollen, spores, paper, tobacco, starch grains, seeds, plant hairs, marijuana, spices, fragments of bricks, concrete, plaster and chalk, diatoms, metal flakes and shavings, rust fragments, blood, skin, bone, various tissues, insects.

table. The room in which the examination takes place should be dust free. If this is not possible, the room should be kept as clean as possible and be situated in a low-traffic area of the laboratory. Adequate tabletop space for laying out the items of physical evidence should be available. Each item of physical evidence is observed first visually and then with a stereomicroscope. After location and documentation (sketching, photographing), all visible traces can be removed by hand with forceps or a needle.

Next, the item of physical evidence should be processed with some sort of clear, transparent tape, as first suggested by Frei-Sulzer.[5]

Finally, when necessary, the item can be vacuumed for dust traces. The vacuum sweeping strap described by Kirk can be used.[6] Although efficient, vacuuming has many disadvantages. It should be used only when absolutely necessary, and only after visual and taping procedures have been conducted.[7–9] Once collected, the sweepings are preserved in paper containers (boxes or paper folds) for storage until examination or for future reference.

Preliminary Examination

Each specimen of dust should be examined with a stereomicroscope for evaluation and sorting. A data sheet such as that in Figure 1 should be prepared for each dust specimen. If the specimen appears to be homogeneous, a representative sample is mounted on a microscope slide in Cargille's Melt Mount with a refractive index (RI) for the N_D of 1.539 at 25 °C. The dust specimen is then covered with a cover glass. The Melt Mount, which must be heated to 60–70 °C, can be applied with a glass rod, eye dropper, or in stick form.[10] Prior to mounting, the specimen should be teased with two needles to loosen the fibers and debris composing the dust. A representative sample of a heterogeneous dust specimen can be mounted in the same manner. After mounting, the specimen is examined with a polarized light microscope (PLM) for characterization and identification.

The data collected in Figure 1 can be used for the initial classification of the materials(s) composing the dust specimen and to guide the examiner to the appropriate identification procedure or scheme. The use of Melt Mount 1.539 and the PLM for the characterization and identification of the trace materials commonly found in dust specimens is briefly discussed here; a comprehensive discussion can be found in the *Forensic Science Handbook*.[11] The results obtained with these methods should be confirmed with other methods of analysis.

Data Sheet

I. Morphology
Fibrous ______ Particulate ______ Both ______

II. Homogeneity
Homogeneous: Yes ______ No ______
Fiber: Separate ______ Cluster ______ Both ______
Particle: Separate ______ Cluster ______ Both ______
Heterogeneous: Yes ______ No ______
Aggregate of both primary forms: Yes ______ No ______
No. of possible fiber types ____________
No. of possible particle types ____________

III. Initial Classification
Shape: __________________ Sketch: ________________________

If fiber:
Hair: Human __________________ Animal __________________
Synthetic fiber ____
Vegetable fiber ____
Other fiber ____
Other fiber ____

If particles:
Mineral grain __________________ Glass chip _________________
Other particulate ______________________________________

Figure 1. Data sheet for the identification of the materials commonly found in dust.

Identification Methods

Human Hair

Human hair occurs in dust in many forms. First, complete hairs consisting of an intact root end, middle portion, and tip, originating from various parts of the human body, are shed on a daily basis. These hairs often become airborne for short periods and eventually collect in the dust of a given environment or locality. Next, complete human hairs, or hair fragments, can accumulate in dust by grooming practices (such as combing) and by forcible means (such as pulling or cutting). Finally, hair that has been burned can become airborne and thus finds its way into the dust of a given location.

The identification of human hair is based on its physical morphology. The identifying characteristics of human hair are readily observed when the hair is mounted in Melt Mount 1.539.[12] Figure 2 shows the three primary anatomical regions of hair used in species identification: the cuticle, the outermost layer of hair that is composed of several layers of overlapping scales; the medulla, the central canal of the hair (can appear present or absent); and the cortex, the primary tissue composing hair. The cortex contains the pigment granules, cortical fusi, and the other morphological features that make up hair. Figure 2 also shows a cast of the dominant scale pattern usually associated with hair of human origin; a temporary scale cast can be prepared in Melt Mount 1.539, and then remounted in Melt Mount 1.539 for further study.

The human hairs encountered in dust specimens usually originate from the head or pubic regions of the body. However, hairs originating from other body areas such as the face and limbs frequently occur. Therefore, examiners should familiarize themselves with the morphology of all types of human hairs. Bisbing and Hicks list the various morphological characteristics used to determine the somatic origin of human hair.[13–14]

The race of the person from whom a questioned hair originated can often be important in forensic investigations. Racial origin is determined on the basis of morphological features. Lists of these features are available in the literature.[14–15]

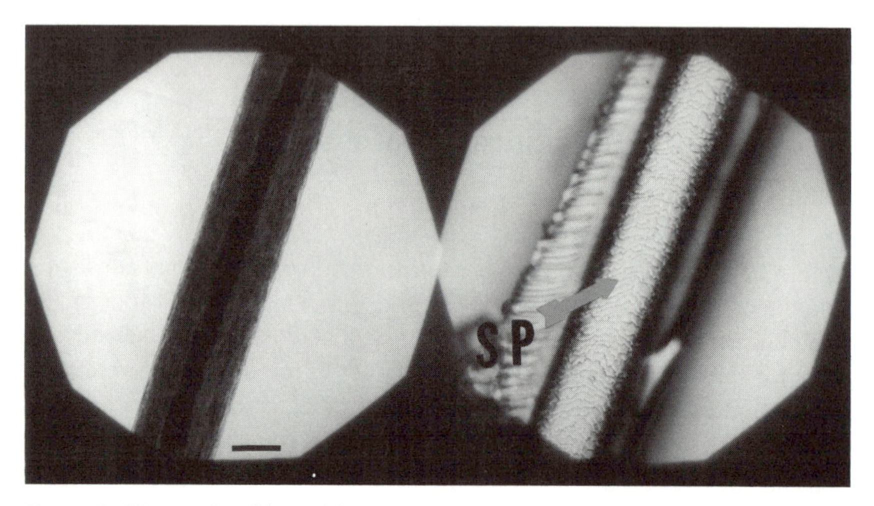

Figure 2. Human head hair. The arrow points to the cuticle impression.

Many protocols for the examination and comparison of human hair can be found in the literature. The following list is an abbreviated protocol for human hair examination.[16–26] The data obtained from such a protocol are used to help establish the somatic and racial origin of a questioned hair, and to show similarities or differences between questioned and known specimens.

Macroscopic characteristics

- Length
- Reflected light color
- Shaft shape
- Texture

Microscopic characteristics (three primary anatomical regions)

- Cuticle
  - Margin
  - Shape
  - Color
  - Thickness
- Cortex
  - Tip shape
  - Transmitted light color
  - Color distribution
  - Cross section
  - Pigment granule shape
  - Pigment distribution
  - Pigment density
  - Shaft diameter
  - Shaft diameter variations
  - Cortical fusi
  - Root structure
  - Proximal end (no root)
  - Growth phase
  - Oddities
  - Damage
  - Substances
- Medullary structure
  - Absent
  - Present
  - Configuration
  - Distribution
  - Thickness
  - Medullary index

Animal Hair

Animal hair can occur in forensic dust specimens both as complete hairs and as fragments. Animal hair collects in dust in much the same manner as does human hair. Many pets shed hair on a daily basis. Hair from grooming pets finds its way into the dust of a given environment. Animal hair from articles of clothing and other textile materials made from animal hair or fur can become airborne and thus be incorporated into the dust of a given location. A scheme to aid in the identification of the various species of animal hair that commonly occur in forensic science casework is briefly described next; I have published an article with complete details.[27]

Complete animal guard hairs present in dust specimens are sorted out during the initial examination. These hairs should be examined visually and with a stereomicroscope. Each hair is sketched, measured, and its reflected light color(s) and color banding noted. The data are recorded. After preliminary examination, the hair's scale pattern is cast in Melt Mount 1.539.[28] Next, a wet mount of the guard hair is prepared in Melt Mount 1.539. The specimen is then examined with a PLM.

The scale cast is examined first. The dominant scale pattern near the root of the hair is noted. Next, the scale pattern(s) from root to tip is scanned and noted. Figure 3 shows examples of the six basic scale patterns. The wet mount is then examined to collect information concerning the specimen's transmitted light color, medullary configuration, and so on. Figure 4 depicts six primary medullary configurations. All the observations are recorded. In order to identify the family of species from which a questioned hair specimen originated, the collected data and the specimen should be

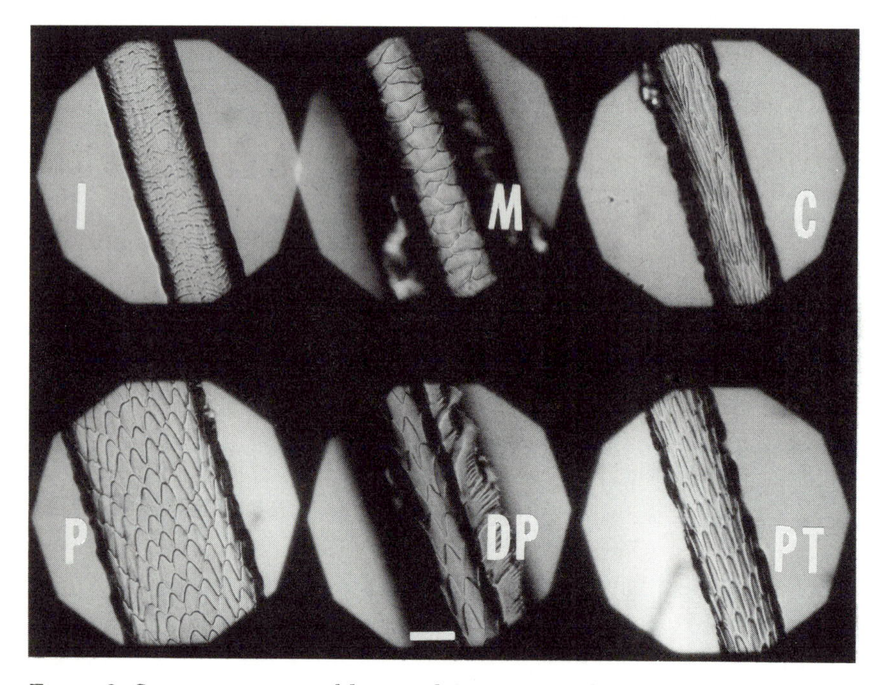

Figure 3. Six primary animal hair scale patterns: imbricated (I), mosaic (M), chevron (C), petal (P), diamond petal (DP), and pectonate (PT).

Figure 4. Six primary animal hair medullary configurations.

compared with the published literature and known reference standards.[27,29–39]

Synthetic Fibers

With the large production today of synthetic fibers for all types of textile products, our environment is literally inundated with fragments of fibers. Dust specimens composed of synthetic fibers rolled together into balls are ubiquitous. Dust balls are formed from the wearing down of textile materials (rugs, clothing, and so forth), as well as the hair from animals and people, natural fibers, and other materials in our environment. Dust balls have been called "urban soil samples" and, like soil samples, often represent the environment(s) in which they are formed.[40] The synthetic fibers entangled in these dust specimens can be identified in the matrix specimen.

The dust specimen is mounted on a microscope slide in Melt Mount 1.539, as previously described. Prior to mounting, the specimen should be teased with two needles to loosen the fibers, hairs, and other debris. After mounting, the preparation is observed with a PLM. The microscopist will observe a variety of fibers. At this point, the examiner uses his or her eyes to single out the fiber to be identified and makes the following observations. Information concerning the fiber's morphology is collected first.

Next, the relative refractive indices (RRI) of the fiber's N| and N⊥ directions, as they compare to the mounting medium's RI (1.539), are obtained by the Beck line method using plane polarized light. In the Beck line method, the fiber's elongated axis is made parallel to the vibrational (preferred) direction of the polarizer. The movement of the Beck line is noted when the microscope's focus is raised (the Beck line moves toward the medium of higher RI under these conditions). The fiber's elongated axis is then made perpendicular to the preferred direction of the polarizer and the movement of the Beck line is noted in this orientation; see Figure 5 for orientation of the fiber and for movement of the Beck line.

Next, the fiber is observed between crossed polars. If the fiber is optically anisotropic, the amount of retardation the fiber exhibits is estimated using an interference chart and the appropriate compensator(s). The fiber's sign of elongation (SE) is determined at this stage of the examination. The fiber's estimated birefringence (EB) is computed using the collected data. Other comparative information about the fiber's appearance (delustering agent, twist, crimp [Figure 6], etc.) and optical properties (degree of relief

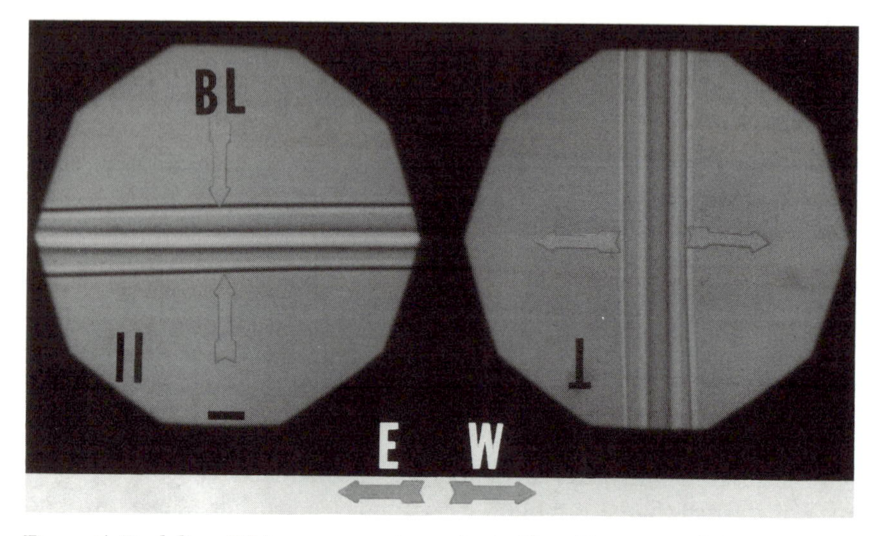

Figure 5. Beck line (BL) movement in synthetic fiber; E is east and W is west.

and so forth) is collected. All the data are recorded in the examiner's notes or on a fiber data sheet. A sample data sheet is shown in Figure 7.

To determine the generic class of a questioned fiber, the data collected in Figure 7 are compared to known published data and to known standards.[11,41–49] Each fiber in the dust specimen is identified in this manner.

Minerals, Glass, and Miscellaneous Substances

McCrone has advocated that a variety of microscopic particles can be identified on the basis of their characteristic morphologies and some simple optical properties.[50] Mineral grains, glass fragments, diatoms, vegetable fibers, wood fibers, cordage fibers, paint flakes, starch grains, feathers, and insect parts are just a few of the substances that can be identified in this manner and that frequently occur in forensic dust specimens. A few of these substances are shown in Figures 8–12 as they appear in Melt Mount 1.539.

The mineral grains, glass fragments, and miscellaneous substances often encountered in forensic dust specimens usually originate from the soil located in the surrounding region, or from some other sources in the environment such as vegetation, human and animal activity, and glass containers. When a dust specimen is mounted in Melt Mount 1.539 and studied with a PLM as previously described, tiny fragments of these types of materials are often observed. Just as hairs and fibers can be identified on the basis of their morphological appearance and optical properties, so can these materials.

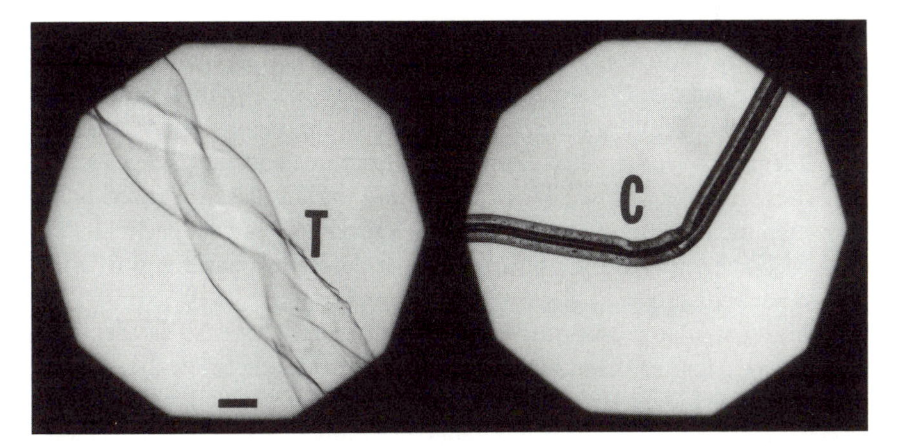

Figure 6. Two forms of synthetic fiber treatment: twisted (T) and crimped (C).

Fiber Data Sheet

Morphology
Longitudinal: Smooth Striated Irregular Other ____________
Cross sectional shape: ______________________
Diameter or lobe(s) thickness in μm: ________________

Optical data
Relative refractive indices—relative to medium (1.539)
N parallel (N||) above below equal
N perpendicular (N⊥) above below equal

Crossed polars: Isotropic Anisotropic
Estimated retardation in nanometers (nm): ______________
(Interference colors)
Estimated birefringence: ________________
Sign of elongation: ________________

Comparative information
Color: ___________ Dyed Undyed
Delustering agent: Bright Semi-dull Dull
Treatment: Crimped Twisted Other ________
Degree of relief: Low Medium High

Other information: ____

Figure 7. A data sheet with spaces for all the information necessary for the classification of synthetic fibers.

Two of the most commonly occurring minerals and mineral-like materials in dust specimens are grains of quartz and glass fragments. These substances are chemically alike (composed of SiO_2), and have similar morphological features—for example, conchoidal fractures and sharp edges—and thus they appear quite similar when viewed under plane polarized light. However, as Miller points out, these two materials can be easily distinguished by the appearance of interference colors in quartz grains when viewed between crossed polars (refer to Figure 8).[51]

Other commonly occurring minerals and related materials are depicted in Figure 9. Most of these substances are easily identified on the basis of their morphological appearance and by a quick determination of some of their optical properties (degree of relief, birefringence, interference colors, and so forth). An eminently good article by Graves on soil mineralogy has proven valuable for characterizing mineral grains.[52]

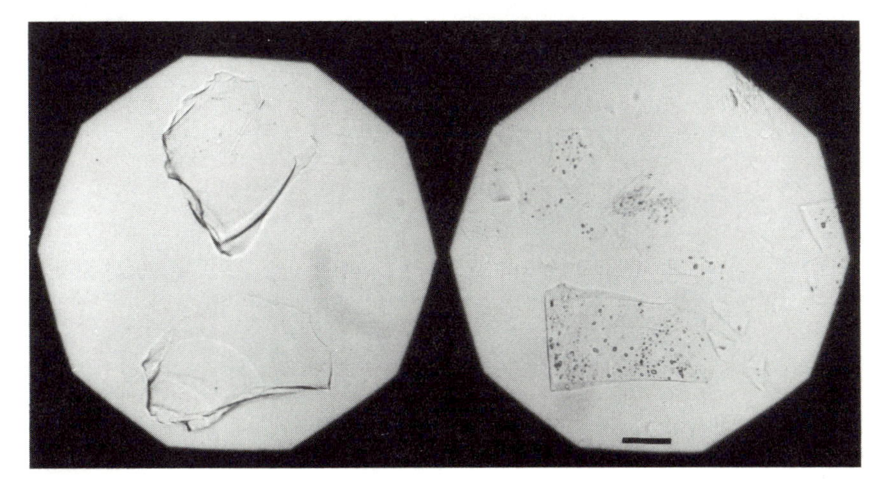

Figure 8. Specimens of glass (left) and quartz (right) fragments.

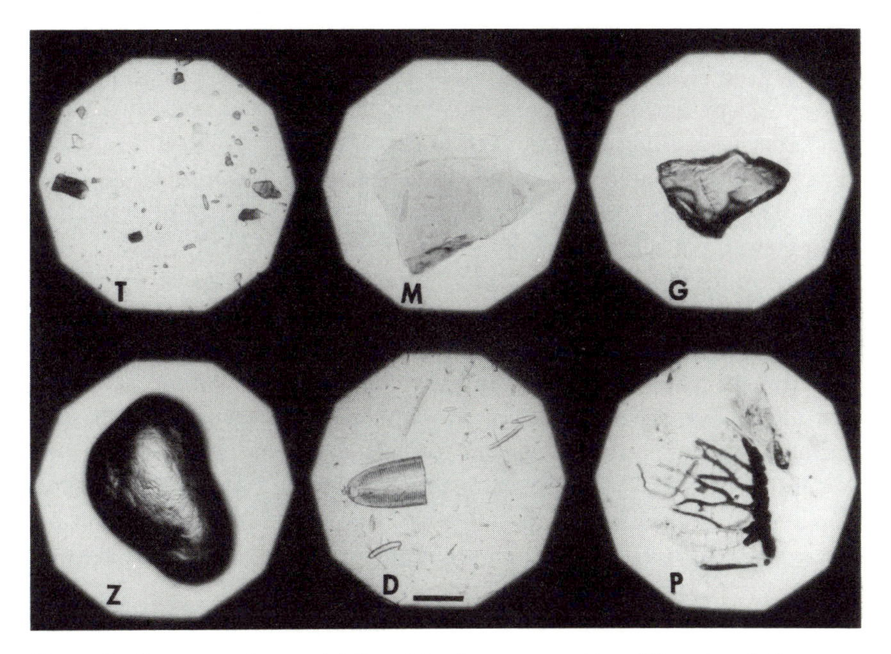

Figure 9. Six types of material found in dust: tourmaline (T), mica (M), garnet (G), zircon (Z), diatom (D), and peat (P).

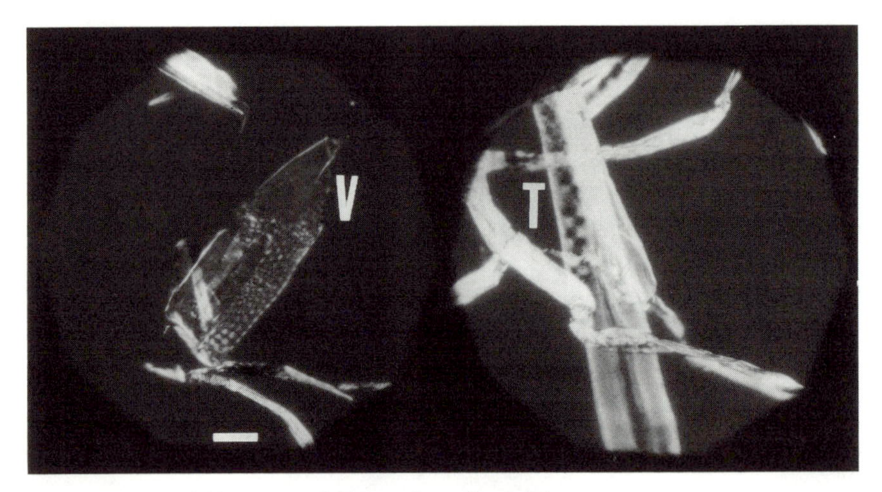

Figure 10. Wood fibers: vessel (V) and tracheid (T).

Wood fibers that originate from paper products and sawdust are ubiquitous in our environment and are consequently found in dust specimens. Wood (paper) fibers can be identified on the basis of their microscopic morphology (see Figure 10). Atlases such as those prepared by Parham and Gray[53] and by Coyote[54] are useful when attempting to identify a wood fiber's species of origin. Cotton fibers are also abundant in our surroundings and thus are seen often in forensic dust specimens. Cotton fibers are easily recognized by their characteristic morphology and their lack of extinction when viewed between crossed polars.[55] Other vegetable fibers commonly observed in forensic dust specimens are tobacco, marijuana, sisal, manila, flax, and ramie (see Figures 11 and 12). An excellent test by Catling and Grayson concerning the identification of vegetable fibers in the forensic laboratory is invaluable when identifying some of these substances.[56] Their book also contains a wealth of information on the identifying features of several other commonly used vegetable fibers.

Grains of starch are seen on a routine basis in casework. Common sources are food products, surgical gloves, and baby powder. Most types of starch can be identified by the shape and size of their grain(s), as shown and described in the literature.[57]

Feathers found in dust specimens usually originate from domesticated birds that are raised for use as food, such as ducks, chickens, and turkeys, or those that are commonly found in our environment, such as pigeons. Feathers from domesticated birds are frequently used as fillers jackets, coats, pillows, and other items. Consequently, they find their way into the dust

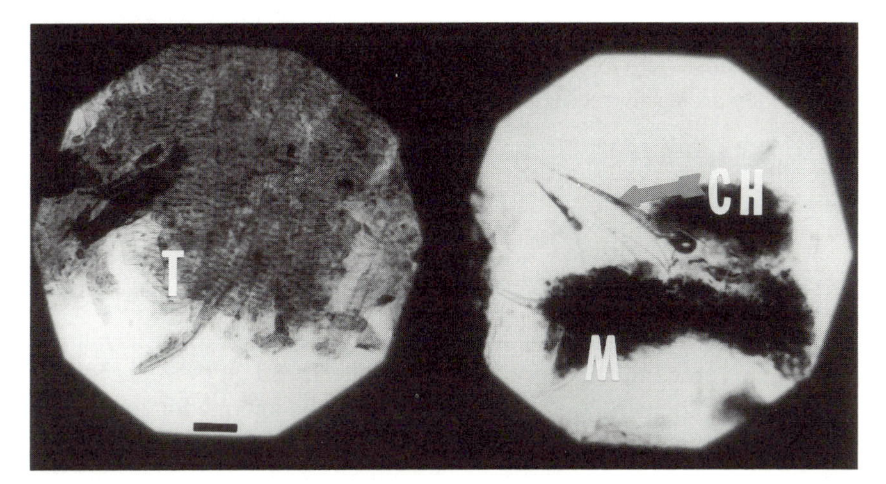

Figure 11. Plant material: tobacco (left) and cystoletic hair (right).

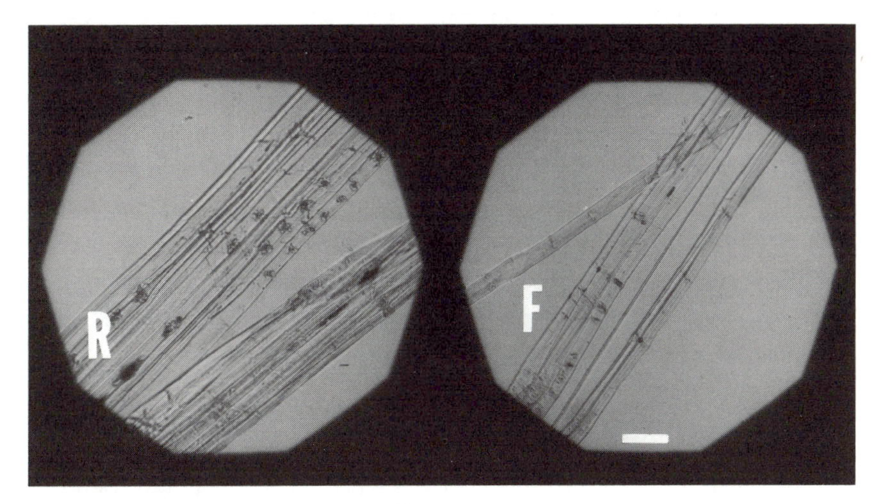

Figure 12. Ramie (R) and flax (F) fibers.

found in many places. Feathers can be identified morphologically. One can usually determine the bird family a feather originated from on the basis of the structure of its down.[58]

Finally, any microscopist who wishes to identify the minerals or other miscellaneous materials that commonly occur in dust specimens should first have a working knowledge of the minerals found in the geographic region served by the laboratory, as well as common building materials, common inorganic salts, various types of glass, foodstuffs, vegetation, and so forth. A set of standards for these materials mounted in Melt Mount 1.539 should also be available. The microscopist should thoroughly study these materials so that they are easily recognized before attempting any identifications in casework.

Dust Comparision

Once the trace evidential contents of any questioned and known dust specimens have been identified, the information can be compiled on tabulation sheets such as the one in Figure 13. The data recorded on these sheets will make the final comparison and interpretation of the information much easier. The data compiled on the tabulation sheet can easily be adapted to form a computer database that could be interpreted by the use of artificial intelligence. Therefore, a tabulation sheet would be prepared for each dust specimen. Finally, the tabulation sheet can be useful when preparing or presenting court exhibits or testimony.

Case Studies

Following are studies of three criminal cases I have worked on in which the microscopic study of dust traces as described in this chapter played a vital role. The cases are adapted by permission of the American Society for Testing and Materials from N. Petraco, "Trace Evidence—The Invisible Witness," *Journal of Forensic Sciences*, **1986,** *31* (9) 321.

Case 1

On a cold winter day in February 1978, the body of a female was found in the alleyway of an east Harlem tenement. In close proximity was a California florist flower box and a plastic liner. The decedent was identified as a member of a well-known church. She was known to have been selling church literature in the buildings that surrounded the alley where her body was discovered. The detectives investigating the case forwarded the flower box, plastic liner, and the decedent's clothing to the forensic laboratory. On the box and liner were found tan wool fibers, red acrylic fibers, and navy blue wool fibers (all identified by polarized light microscopy). The three types of questioned fibers were compared microscopically with the decedent's clothing. All three were found to be consistent, in all respects, to the

Dust Specimen Comparison Sheet

Dust specimen source: Questioned ________ or Known ________

Human hair: Yes ____ No ____

Racial origin: Caucasian ___ Mongoloid ___ Negroid ___ Mixed ___
Somatic origin: Head ____ Pubic ____ Other ____________________
No. of hair types ____ Race(s) ____ Body area(s) ______________

Animal Hair: Yes ____ No ____
Guard hair ____ Fur ____ Other ____________________
Species of origin: ____________________________________
No. of different species: ____________________________________

Synthetic fibers: Yes ___ No ____
Generic class: ____________________________________
No. of different types of each generic class:
Acetate ____ Triacetate ____ Acrylic ____ Aramid ____
Modacrylic ____ Polyamide ____ Polyester ____ Olefin ____
Rayon ____ Other ____________________________________

Vegetable fiber or /matter: Yes ____ No ____
Type present:
Cotton ____ Ramie ____ Sisal ____ Flax ____
Other ____________________________________

Minerals, glass, and related materials: Yes ____ No ____
Glass ____ Mineral ____ Other ____________________________
Type present: ____________________________________

Known and Questioned Similar ____ Dissimilar ____ Both ____

No. of similar materials in questioned & known dust ________________

No. of dissimilar materials in questioned & known dust ______________

Figure 13. A sheet for the comparison of dust specimens.

textile fibers composing the decedent's clothing (tan wool overcoat, navy blue wool–polyester blend slacks, and red acrylic sweater), thereby associating the woman with the flower box and liner. In addition, light blue nylon rug fibers and several brown rabbit hairs were found on the box and liner. Similar light blue nylon rug fibers and rabbit hairs, as well as red nylon rug fibers, were found on the decedent's tan wool overcoat. Neither the rabbit hairs nor the nylon rug fibers could be associated with the victim's environment (her clothing or residence).

All this information was conveyed to the field investigators. Upon further inquiry in the neighborhood, the investigating officers learned the identity of a man who had, the day after the body was discovered, sold a full-length, brown rabbit coat to a local man. The investigators obtained the rabbit coat from the purchaser. The hair composing the coat was compared microscopically to the questioned rabbit hairs found on the victim's wool coat and the flower box liner. The specimens of questioned rabbit hair were found to be consistent in all physical and microscopic characteristics to the rabbit hair composing the suspect's coat, (see Figure 14). Armed with this information, the police had probable cause to obtain a search warrant for the suspect's apartment.

In the suspect's apartment two rugs were found. One was light blue and the other red; both rugs were nylon. Samples of each rug were taken by the crime scene unit and forwarded to the forensic laboratory for comparison with the questioned rug fibers found on the victim's clothing and on the flower box and plastic liner. Both the questioned and known rug fibers were found to be consistent in all respects. The presence of light blue nylon rug

Figure 14. Comparison of rabbit hair, questioned (left) and known (right).

fibers, red nylon rug fibers, and brown rabbit hairs on the flower box, plastic liner, and woman's clothing enabled me to associate the woman and the flower box and liner found in the alleyway with the suspect and his apartment. The various associations made possible by the trace evidence are shown in Table II.

Further inquiry concerning the suspect was made in the neighborhood by the investigating officers. A witness was located who stated he saw the suspect carrying a large California flower box a day or two before the body was discovered.

From the evidence it was theorized that the woman was killed in the suspect's apartment, placed in the flower box, brought up to the roof of the building in which the defendant resided, and thrown off the building into the alley below. On the basis of all of this evidence, the suspect was arrested, indicted, and tried for murder in the second degree. After two trials, at which I gave three days of testimony about the trace evidence, the defendant was found guilty of murder in the second degree and sentenced to life imprisonment.

This case demonstrates the kinds of associations that can be made by trace dust evidence between people, places, and things. Another form of information that can often be obtained from trace evidence is describing the occupation or environment of the principals in a case. The next case is a good example of how this information can be useful.

Case 2

In June 1978, the body of a woman in her twenties, was found in a parking lot in midtown Manhattan. Black fibrous material was found in the victim's

Table II. Associations Between the People, Places, and Things in Case 1.

Association made from	*People*		*Places*		*Things*	
	Victim	*Suspect*	*Alley*	*Apartment*	*Liner*	*Box*
Textile fibers from victim's clothes					x	x
Trace fibers from victim's clothes		x		x		
Suspect's coat (rabbit hair)	x		x		x	
Suspect's apartment (nylon rugs)	x		x		x	

hands by the medical examiner. Microscopic examination of the black fibers revealed a blend of synthetic fibers that consisted of 80% modacrylic fibers, 15% acrylic fibers, and 5% polyester fibers. A search of the victim's husband's van, which was found in New Jersey, produced similar-looking tufts of black fibrous material. Although the van's interior had been recently cleaned and stripped of its carpeting, black fibers were found on a wooden plant holder that the husband used in his floral business and in the samples of vacuum sweepings collected from the rear of the van.

Polarized light microscopy analysis of the black fibers found in the van revealed the presence of a blend of synthetic fibers: 80% modacrylic, 15% acrylic, and 5% polyester. The comparison of the black synthetic fibers found in the victim's hands and the black synthetic fibers found in the van disclosed them to be consistent in all respects.

During the investigation, a question arose concerning the victim's husband's previous occupation. Although he was now employed in the floral industry, it was believed that he was at one time a building contractor. The investigator wanted to know if the van had been used to transport building materials. Polarized light microscopic examination and analysis of the dust removed from the van by vacuuming disclosed the presence of several trace materials that could be associated with the building industry: aluminum–brass chips, dried adhesive compound, fiberglass insulation and resin, gypsum fragments (plaster), iron shavings and filings, mica chips, plate glass fragments, red brick fragments, and wood chips and shavings (sawdust).

Although the listed items do not conclusively prove the occupation of the van's owner or user, they at least provide a strong indication. At the trial, both the black fibers and particulate matter were used to implicate the woman's husband and his accomplice in her death.

Occasionally, unequivocal association between people, places, and things can be made by trace evidential materials. The following case is an example of such an occasion.

Case 3

In the early morning hours of April 12, 1982, atop a lonely roof garage on the west side of Manhattan, three men were found murdered. Each man had been shot once in the back of the head. A light-colored van was seen speeding away from the scene. Hours later, in a secluded alley-street on the lower east side of Manhattan, two dog-walkers found the body of a fully clothed woman lying face down. The woman had been killed in the same manner as the men on the roof garage. The condition of the woman's body and other evidence made it apparent that she had been shot at the garage and then transported to the alley.

An eyewitness to the incident stated that he saw a man shoot a woman and place her in a light-colored van. The gunman then chased down the three men who were coming to the woman's aid and shot each of them. Days later, the prime suspect to the killings was arrested in Kentucky in a black van.

Numerous items of evidence (over 100) were collected from the van and forwarded to the New York City police laboratory for examination. Among the items forwarded were three sets of vacuum sweepings from the van's interior. An autopsy of the woman produced several items of trace evidence that were forwarded to me for microscopic examination. The woman's clothing also was received for trace analysis.

A prime question that arose during the investigation was, could the woman's body, which had been placed in a light-colored van at the garage and later left in an alley on the lower east side, be associated with the black van recovered more than 600 miles away from the scene? Microscopic analysis and comparison of the trace evidential materials found on the victim and inside the van made this association possible. Table III lists all the items of similar trace materials that the victim and the van had in common.

Microscopic comparisons of the questioned human head hair present on the victim's clothing were made with known samples. Ten of the brown and gray Caucasian head hairs from the victim's blazer were consistent in microscopic characteristics to the defendant's known head hair sample. One chemically treated head hair found on the victim was consistent in microscopic characteristics to the known head hair sample obtained from the defendant's wife. One forcibly removed brown Caucasian head hair that was found on the rear door of the van's interior by the Kentucky state police was found to be consistent in all characteristics with the decedent's known head hair sample.

Microscopic comparisons of the white and brown-and-white colored dog hair from the victim's clothing and the van's interior were made with known samples of dog hair obtained from a dog owned by the defendant's nephew, the van's previous owner. The questioned dog hairs were found to be consistent with the hair from the nephew's dog.

The white seed that was recovered from the victim's mouth by the medical examiner and the white seed I found in the van's sweepings were forwarded to an internationally known botanist for identification and comparison. During the trial, the botanist testified that the two seeds were identical in all respects and that although he could not identify the seeds, both

Table III. Items of Similar Trace Evidence Recovered from Both the Victim and the Van Interior.

	Source	
Trace Evidence	*Victim*	*Van*
White seed	Mouth	Sweepings
Paint chips, gray-metallic–black	Hair and wool blazer	Sweepings and floor
Saw dust	Hair, blazer, and sheet	Sweepings and misc. items.
Glass Fragments		
Clear	Wool blazer and sheet	Sweepings and misc. items
Amber		
Green		
Cellophane	Wool blazer	Floor
Urethane foam	Wool blazer	Sweepings and misc. items and foam mattress
Blue olefin plastic	Skirt	Floor
Dog hair		
Brown and white	Wool blazer	Sweepings and misc. items
White		
Human hair		
Brown	Wool blazer	Hair brush, sweepings, and misc. items.
Gray		

were from the same species of plant, if not the same plant, which was probably a rare wildflower.

Sixteen gray-metallic–black paint chips from the victim and her clothing were compared to the gray-metallic–black paint removed from the van. Samples from the questioned and known sources were examined and compared by microscopic, chemical, and instrumental means. All of the paint specimens from the van and from the victim were found to be similar in all respects.

The remaining items of trace evidence from the victim and the van were examined and compared microscopically and, where necessary, by chemical and instrumental methods. Each of the remaining types of trace evidence from the victim was found to be similar to its counterpart from the van.

Blue and black flakes of acrylic paint were found in the van's sweepings and on the suspect's sneakers. No blue or black paint flakes were found on the victim or her clothing. During a crime scene search of the defendant's residence in New Jersey, a large quantity of blue and black acrylic paint was found in the garage. It was apparent from the evidence in the defendant's garage that a large rectangular object had recently been painted with blue and black paint. The blue and black paint flakes from all the sources and known blue (undercoat) and black (topcoat) paint from the van were compared by microscopic, chemical, and instrumental means. All the samples of paint were found to be consistent in every respect.

At the trial, I gave two days of testimony concerning the collection, examination, identification, and comparison of the trace evidence from the victim and the van. When questioned about the source of the trace evidence found on the victim and her clothing, I stated unequivocally that the trace evidence on the victim was from the defendant's van. On the basis of this evidence and other circumstantial evidence, the defendant was found guilty of all charges and sentenced to 100 years in prison.

Conclusion

The importance of the analysis of forensic dust specimens in criminal cases cannot be overemphasized. As shown by the case studies, the information obtained can be used to help reconstruct the incident, describe the occupation(s) of the principals in the case, and describe the scene(s) or location(s), as well as to associate the people, places, and things involved in the event. Dust is a powerful witness indeed, and one that is sorely needed by our criminal justice system.

References

1. Gross, H. *Criminal Investigation,* adapted from *System Der Kriminalistik,* by J. C. Adams; Sweet and Maxwell: London, England, 1924; pp 144–147.
2. Locard, E. *Am. J. Police Sci. 1* **1930,** *Part I,* 276–298; *Part II,* 401–418; *Part III,* 496–514.
3. Nickolls, L. C. In *Methods of Forensic Science;* Lundquist, F., Ed.; Interscience: New York, 1962; Vol. 1, pp 335–371.
4. Locard, E. *Revue Internationale de Criminalistique* **1929,** *14 Juillet,* 176–249.
5. Frei-Sulzer, M. *Kriminalistik* **1951**, *No. 19/20,* 190–194.
6. Kirk, P. L. *J. Criminal Law, Criminology and Police Sci.* **1949–1950**, *40,* 362–369.
7. DeForest, P. R.; Gaensslen, R. E.; Lee, H. C. *Forensic Science: An Introduction to Criminalistics;* McGraw: New York, 1983.
8. Petraco, N. *J. Forensic Sci.* **1985,** *30,* 486.
9. Palenik, S. In *Forensic Science Handbook;* Saferstein, R., Ed.; Prentice-Hall: Englewood Cliffs, NJ, 1988; Vol. II, pp 165–167.
10. DeForest, P. R.; Ryan, S.; Petraco, N. *The Microscope* **1987**, *35,* 261–266.

11. Petraco, N.; DeForest, P. R. In *Forensic Science Handbook;* Saferstein, R., Ed.; Prentice-Hall: Englewood Cliffs, NJ: Vol. III, 1993.
12. DeForest, P. R.; Shankles, B.; Sacher, R. L.; Petraco, N. *The Microscope* **1987**, *35*, 249–259.
13. Bisbing, R. E. In *Forensic Science Handbook;* Saferstein, R., Ed.; Prentice-Hall: Englewood Cliffs, NJ, 1982; pp 201–202.
14. Hicks, J. W. *Microscopy of Hair;* U.S. Government Printing Office: Washington, DC, 1977; Issue 2, pp 7–10.
15. Bisbing, R. E. In *Forensic Science Handbook;* Saferstein, R., Ed.; Prentice-Hall: Englewood Cliffs, NJ, 1982; p 201.
16. Glaister, J. *A Study of Hairs and Wools Belonging to the Mammalian Group of Animals, Including a Special Study of Human Hair, Considered from Medico-Legal Aspects;* MISR: Cairo, Egypt, 1931; p 155.
17. Smith, S.; Glaister, J. *Recent Advances in Forensic Medicine,* 2nd ed.; Blakiston's Son: Philadelphia, PA, 1939; pp 118–124.
18. Gaudette, B. D.; Keeping, E. S. *J. Forensic Sci.* **1974**, *10*, 601–602.
19. Gaudette, B. D. *J. Forensic Sci.* **1976,** *21*, 515–516.
20. Hicks, J. W. *Microscopy of Hair;* U.S. Government Printing Office: Washington, DC, 1977; Issue 2, pp 6–27.
21. Bisbing, R. E. In *Forensic Science Handbook;* Saferstein, R., Ed.; Prentice-Hall: Englewood Cliffs, NJ, 1982; pp 199–205.
22. McCrone, W. C. In *The Particle Atlas;* McCrone, W. C.; Delly, J. G.; Pelanik, S. J., Eds.; Ann Arbor Science: Ann Arbor, MI, 1979; Vol. 5, p 1383.
23. Shaffer, S. A. *The Microscope* **1982,** *30*, 151–161.
24. Strauss, M. A. T. *The Microscope* **1983,** *3331*, 15–29.
25. Gaudette, B. D. *Crime Laboratory Digest* **1985,** *12*, 44–59.
26. Lee, H. C.; Deforest, P. R. In *Forensic Sciences;* Wecht, C. H., Ed.; Matthew Bender: New York, 1987; Vol. 3, pp 37A–8 & 9.
27. Petraco, N. *The Microscope* **1987,** *35*, 83–92.
28. Petraco, N. *The Microscope* **1986**, *34*, 341–345.
29. Glaister, J. *A Study of Hairs and Wools Belonging to the Mammalian Group of Animals, Including a Special Study of Human Hair, Considered from Medico-Legal Aspects;* MISR: Cairo, Egypt, 1931.
30. Smith, S.; Glaister, J. *Recent Advances in Forensic Medicine,* 2nd Ed.; Blakiston's Son: Philadelphia, PA, 1939; pp 86–124.
31. Kirk, P. L. *Crime Investigation;* Interscience: New York, 1953; pp 152–175.
32. Hicks, J. W. *Microscopy of Hair;* U.S. Government Printing Office: Washington, DC, 1977; Issue 2, pp 28–40.
33. Sato, H.; Yoshino, M.; Seta, S. *Reports of National Research Institute of Police Science* **1980,** *33(1)*, 1–16.
34. Wildman, A. B. *Microscopy of Animal Textile Fibres;* WIRA: Leeds; p 19.
35. Adoryan, A. S.; Kolenosky, G. B. *A Manual for the Identification of Hairs of Selected Ontario Mammals;* Research Report Wildlife, No. 90; Dept. of Lands and Forests: Ontario, 1969.
36. Moore, T. D.; Spence, L. E.; Dugnolle, C. E.; Hepworth, W. G. *Identification of the Dorsal Guard Hairs of Some Mammals of Wyoming;* State of Wyoming: Cheyenne, WY, 1974.

37. Brunner, H.; Coman, B. J. *The Identification of Mammalian Hair;* Iukata: Melbourne, 1974.
38. Appleyard, H. M. *Guide to the Identification of Animal Fibres,* 2nd Ed.; WIRA: Leeds, 1978.
39. McCrone, W. C. In *The Particle Atlas;* McCrone, W. C.; Delly, J. G.; Pelanik, S. J., Eds.; Ann Arbor Science: Ann Arbor, MI, 1979; Vol. 5, pp 1383–1384.
40. Petraco, N. *J. Forensic Sci.* **1987,** *32,* 768.
41. *The Particle Atlas,* 2nd Ed.; McCrone, W. C.; Delly, J. G., Eds.; Ann Arbor Science: Ann Arbor, MI; Vol. 1, 1973; Vol. 1, 1974; Vol. 2, 1977; Vol. 4, 1977; McCrone, W. C.; Delly, J. G.; Palenik, S. J., Eds.; Vol. 5, 1979.
42. Gaudette, B. D. In *Forensic Science Handbook;* Saferstein, R., Ed.; Prentice-Hall: Englewood Cliffs, NJ, 1988; Vol. II, pp 209–272.
43. O'Neill, M. E. *J. of the American Institute of Criminal Law and Criminology* **1934,** *25,* 835–842.
44. Longhetti, A.; Roche, G. W. *J. Forensic Sci.* **1958,** *3,* 303–329.
45. Rouen, R. A.; Reeve, V. C. *J. Forensic Sci.* **1970,** *15,* 410–432.
46. Forline, L.; McCrone, W. C. *The Microscope* **1971,** *19,* 243–254.
47. *Identification of Textile Materials,* 7th Ed.; The Textile Institute: Manchester, England, 1975.
48. National Bureau of Standards. *Reference Collection of Synthetic Fibers*; U.S. Dept. of Commerce: McLean, VA, 1984.
49. Petraco, N.; Deforest, P. R.; Harris, H. *J. Forensic Sci.* **1980,** *25,* 5571–582.
50. *The Particle Atlas,* 2nd ed.; McCrone, W. C.; Delly, J. G., Eds.; Ann Arbor Science: Ann Arbor, MI., 1973; Vol. 1, Introduction.
51. Miller, E. T. *Forensic Science Handbook;* Saferstein, R., Ed.; Prentice-Hall: Englewood Cliffs, NJ, 1982; p 154.
52. Graves, W. J. *J. Forensic Sci.* **1979,** *24,* 331–337.
53. Parham, R. A.; Gray, R. L. *The Practical Identification of Wood Pulp Fibers;* Tappi: Atlanta, GA, 1982.
54. *Papermaking Fibers: A Photomicrographic Atlas;* Coyote, W. A., Ed.; Syracuse University: Syracuse, NY, 1980.
55. *The Particle Atlas,* 2nd Ed.; McCrone, W. C.; Delly, J. G., Eds.; Ann Arbor Science: Ann Arbor, MI, 1974; Vol. 2, pp 352–3.
56. Catling, D. M.; Grayson, J. E. *Identification of Vegetable Fibres;* Chapman Hall: London, 1982.
57. *The Particle Atlas,* 2nd Ed.; McCrone, W. C.; Delly, J. G., Eds.; Ann Arbor Science: Ann Arbor, MI, 1974; Vol. 2, pp 398–462.
58. Metropolitan Police Forensic Science Laboratory. *Biology Methods Manual*; Commissioner of the Metropolis: London, 1978; 6-9 & 6-11.

5

Laboratory Examination of Arson Evidence

Charles R. Midkiff, Jr.

Arson is an ancient crime but one that is difficult to investigate and prosecute even with today's technology. Traditionally, arson has been defined as "willful and malicious burning". There are two elements to a successful prosecution for arson: demonstrating that there was a fire and proving that it was intentionally and maliciously set. Only after each is done can an individual be convicted of the crime. Although both elements also are required to prosecute other crimes, showing the nonaccidental cause of death in, for example, a homicide is usually not difficult. In a suspected arson, that a fire took place is usually uncontested, but other aspects of the crime may pose problems. For instance, in the destruction typical of a fire, evidence of arson may be lost or difficult to locate. Nevertheless, it must be shown in court that the fire was intentionally started.

Circumstances of Arson and Arson Prosecution

With extensive damage at the scene, establishing where and how the fire began—origin and cause determination—challenges even the experienced fire investigator. The point of origin is important in fire investigation because where the fire started is the most likely location for evidence of its cause. In suspicious fires, the origin is also the most productive location for evidence proving that it was incendiary and not accidental. Setting a fire does not in itself constitute arson. For example, a farmer burning brush does not commit arson because he willfully starts the fire, even when it gets out of control and damages the property of someone else. The farmer may have violated an ordinance against open burning or may be found negligent in allowing the fire to get out of control, but unless he intended it to cause damage to his neighbor he is not guilty of arson. To establish the crime of arson, malice must be proven beyond a reasonable doubt. The fire must be shown to have been set with criminal intent—for revenge, financial gain, covering evidence of another crime, or another reason. Only after a reasonable showing that arson was committed can the prosecution introduce evidence to link the defendant with criminal responsibility for the fire.

Generally, arson is a crime committed in secret, without witnesses. Often the fire destroys the evidence upon which its subsequent investigation must rely. For these reasons, prosecutors have long considered arson to be among the most difficult crimes to prove at trial, and in the past they were reluctant to undertake cases without direct or eyewitness evidence. At one time, fire investigators facetiously contended that when they approached prosecutors with a prospective arson case, the first question was, "Did anyone see him strike the match?"— not an auspicious beginning to a successful criminal prosecution.

Because sufficient direct evidence is often unavailable, cases may rely heavily on indirect or circumstantial evidence, of which physical evidence is a major part. Examination by the crime laboratory of material collected at the fire scene plays an increasingly significant role in arson investigations.

To ensure effective destruction, an accelerant is often used in arson to increase the rate of spread and intensity of the fire. Detection of an accelerant in the debris at the scene not only suggests a nonaccidental cause but also implies malice in starting the fire. Although a variety of materials are usable as accelerants because they are readily available and burn well, the most common accelerants are flammable liquids, typically commercial petroleum-derived products such as gasoline, kerosene, fuel oil, charcoal lighter, paint thinner, lighter fluid, and the like. Other flammable liquids include alcohols, paint removers, turpentine, and certain solvents and

Table I. Flammable and Combustible Liquid Classification System

Class/Type	*Carbon Number Range*	*Examples*
1. Light petroleum distillates (LPDs)	C4–C8	Petroleum ethers, pocket lighter fuel, some rubber cements, solvents, VM&P naphtha
2. Gasoline	C4–C12	All brands and grades; gasohol, some camping fuels
3. Medium petroleum distillates (MPDs)	C8–C12	Paint thinner, mineral spirits, some charcoal starters, dry cleaning solvents, some torch fuels
4. Kerosene	C9–C16	No.1 fuel oil, Jet-A insect sprays, some charcoal starters and torch fuels
5. Heavy petroleum distillates (HPDs)	C10–C23	No.2 fuel oil, diesel fuel

industrial chemicals, but these are encountered far less frequently. See Table I for a more complete listing by classification.

When a flammable liquid is poured onto a porous material such as a carpet, it rapidly penetrates and is absorbed. After it is ignited, the porous material functions much like a wick in a lamp. Because liquid fuel can burn only at the surface, where there is sufficient oxygen, combustion occurs at the surface. As the vapors at the surface are consumed, more liquid rises, vaporizes, and burns. If the fire is extinguished quickly or there is insufficient air to support combustion, as in a tightly sealed room, some of the original liquid may remain unburned. Because the heat of fire tends to evaporate the more volatile part of the accelerant, the remaining liquid may differ somewhat in composition from the original. Nevertheless, if residual liquid can be recovered even in trace amounts, the accelerant often is identifiable in the laboratory.

During investigation of a fire, samples are collected from the area of origin and other areas showing intense burning and placed in a vapor-tight container. The most satisfactory and widely used container is a new, clean, metal paint can. These cans are sturdy, available in a range of sizes, relatively inexpensive, and easily sealed by the investigator in the field to prevent loss of vapors or evidence contamination.

Although arson prosecutions have involved testimony that an odor of a petroleum product was noted at the scene, that the rate of spread or intensity of the fire was greater than expected from simple burning of the materials present, or that the color of the smoke or flame suggested use of a petroleum product, these observations provide, at best, weak evidence in court. In modern society, plastics are commonplace, and most are produced from petroleum or natural gas; when they are involved in a fire, observations about odor, smoke, or flame are of especially limited reliability. Because plastics may behave in ways similar to flammable liquids, particularly when melted, they cause major problems to both the investigator at the scene and the chemist in the laboratory. Thus it is essential that the laboratory distinguish residual flammable liquids from plastic by-products.

In arson investigations and prosecutions, a showing of the presence of residual accelerant in physical evidence from the fire scene can provide investigative information or support the findings and testimony of the arson investigator. Although products containing flammable liquids are common, a flammable liquid identified in samples from an area where such materials would not normally be present provides powerful evidence that the fire was not accidental but intentionally set. One example: In 1991, a fire at the Happy Land Supper Club in New York City resulted in the deaths of 86 people in the dining area on the second floor, and gasoline was detected in debris from the first-floor entryway to the club. The laboratory findings also supported other evidence that the suspect had thrown gasoline into the first-floor entry hall and ignited it.

Criminal cases can be prosecuted entirely on circumstantial evidence, without a direct showing that the defendant started the fire, but the evidence must exclude any other reasonable explanation. This places a heavy burden on the state in a type of crime difficult to prosecute even with direct or eyewitness testimony of the defendant's involvement. Because the amount of accelerant remaining in the evidence may be extremely small, changed in composition, or obscured by the presence of materials produced during the fire, careful scientific examination of the debris from the fire scene is essential.

During the last several decades, considerable progress has been made in developing and improving techniques for the laboratory examination of fire scene evidence. Much of this progress is attributable to the efforts of dedicated scientists and technicians who often saw, in new developments in instrumentation and techniques, potential for application to the examination of fire debris. These people were the pioneers in the laboratory examination of arson evidence.

Chemists Begin to Help

Vacuum Distillation Techniques

One of the first reported examples of a scientific approach to the recovery of a liquid accelerant from debris collected at a fire scene occurred in 1940. At the time, crime laboratories were few, inadequately staffed with trained chemists, and lacking in analytical equipment. Vincent Hnizda, a chemist with the Ethyl Corporation, was asked by the Detroit police for assistance in an arson–murder case. He used an innovative approach to separate the unburned accelerant from the debris recovered at the scene. He set up a desiccator, normally used as a drying chamber for chemicals, and placed the debris inside. He connected the outlet of the sealed desiccator by tubing to a series of cooled test-tube-type traps (cold traps) and attached a vacuum pump to the last trap in the series to reduce the pressure in the desiccator and cause liquids in the evidence to evaporate. As the vapor passed into the cold traps, it condensed into liquid separate from the bulk debris and suitable for identification.[1] This approach had a lasting impact on arson evidence examination; a descendant of Hnizda's desiccator-vacuum technique is today the most widely accepted separation and concentration technique for examination of debris collected at the scene of a suspicious fire. Although he received little recognition at the time, Hnizda should be recognized as a true pioneer in the field.

In 1952, J. M. Macoun began a laboratory examination of arson evidence by first smelling the sample. This is not always reliable in the presence of other odors, but sometimes it suggests the presence of a petroleum product in the sample. He then placed the debris in a distillation apparatus and distilled it with alcohol and water. The distillate was collected in a calibrated tube known as a burette, so that its volume could be determined and the sample easily removed for further testing. The sample was tested for flammability and its refractive index was determined. (The refractive index of a liquid, a measure of the extent to which light passing through it is bent, is an indicator of its composition.) The specific gravity (weight per unit volume) of the collected liquid also was determined. Water has a specific gravity of 1; most petroleum products have lower specific gravities, depending on their composition. The temperature range over which the collected liquid boiled, which also helps classify petroleum products, also was tested. By comparing the results of his tests to those of common flammable liquids, Macoun could tentatively identify the separated material as a petroleum product and probable accelerant in the fire.[2] Although not definitive by today's standards, such an examination of the physical evidence from a fire scene produced information that could be introduced in court to show the incendiary nature of the fire.

In the 1950s, techniques for the identification of arson accelerants were limited in sensitivity and required that a relatively pure liquid sample be separated from the fire debris and collected for analysis. As a result, efforts were directed toward development of techniques for more effective separation and concentration of residual flammable liquid from the physical evidence prior to its chemical analysis. One analyst used vacuum distillation, similar to that used by Hnizda, to separate from a rug a liquid subsequently identified as gasoline. He then examined a small amount of liquid remaining in a jug recovered near the scene and identified it as gasoline. A sample obtained from a gas station, where it was thought that the suspect bought the gasoline, also was examined. It was reported that this sample was "analyzed and found to be identical to the gasoline contained in the jug found near the scene of the fire as well as the sample extracted from the rug".[3]

Unfortunately, the stated results from these comparisons are probably optimistic. No details were given as to the analytical methods used to make the comparison, but several questions arise as to their effectiveness. First, the gasoline recovered from the rug was exposed to the heat from the fire and almost certainly underwent some evaporation of components with a low boiling range. This alone would make its composition differ from the relatively fresh gasoline in the jug and from that at the gas station. Second, gasoline is a commodity product, and although gasolines from various producers and batches differ, they are quite similar in overall composition. It is unlikely that the instrumental techniques available at the time would have shown minor differences between the samples, making suspect the conclusion that they were "identical". Perhaps it would have been better had the analyst reported that the samples were "indistinguishable", indicating that although no differences were observed, they could not be precluded.

Despite any technical inadequacies in this instance, comparison of a sample from the crime scene with one available to the suspect is a routine request made of the forensic laboratory. Even with the technology available in the modern crime laboratory, it is still a major challenge to effectively compare materials as complex yet similar as commercial petroleum-based flammable liquids. For a variety of reasons, definitive identifications such as brand name or intended use of the product are rarely possible.

Steam Distillation

Another technique for separation of flammable liquids from arson evidence that still has limited use today is steam distillation. In this process, the sample is placed in a glass distillation apparatus, either water or glycerin is added, and steam is passed into the flask, causing the flammable liquid to distill over into another flask where it is collected. In tests using known amounts of gasoline, in one hour the technique recovers about 70% of the original amount added. For heavier products such as dry-cleaning fluid or kerosene, glycerol is more effective.[4] An attraction of steam distillation is that when sufficient residual accelerant is present in the debris, it is recovered as a relatively clean liquid. Following laboratory testing of the liquid for identification, at trial it can be shown to and even smelled by the jury; this is highly effective in convincing the jury of the validity of the laboratory results.

Solvent Extraction

In 1957, Adams reported an improvement on the vacuum distillation approach.[5] This approach, now known as "purge-and-trap", uses a small hand pump to pass air through a container of arson debris while it is heated with an infrared lamp. As the steam does in steam distillation, the air sweeps accelerant vapors from the container into a cold trap where they condense; this improves recovery when only small amounts of residual flammable liquid are present in the debris sample. Because the vapor pressure of heavier products—such as fuel or lubricating oil—is low compared to gasoline or lighter fluid, heavier products are not recovered from debris as vapors as effectively as the lighter products. Extraction from debris with a volatile solvent such as carbon tetrachloride is more effective with these materials.

After extraction, the solvent is evaporated to a small volume to concentrate the oil for analysis. Because the solvent boils at a temperature well below that of oil, no significant loss of the flammable liquid occurs during the concentration step. Solvent extraction, however, often dissolves other materials in the debris, which can interfere in the analysis and complicate interpretation of the results. A similar approach—rinsing nonporous materials, such as glass, with a volatile solvent—is also effective in recovery of flammable liquids.

Heated Headspace Sampling

Popularized by Midkiff and Washington,[6] heated headspace sampling involves oven heating of a can containing the debris and removal with a gas-tight syringe of several milliliters of the air and flammable liquid vapor above the debris. Although fast, simple, and effective for volatile products such as gasoline, headspace sampling is less effective for heavier products such as kerosene or fuel oil. Though it is useful for quickly screening a number of samples, when headspace sampling yields inconclusive results, solvent extraction or another approach is necessary to ensure that a heavier accelerant is not present.

Carbon-Wire Technique

For nearly 20 years, the major methods of sampling arson evidence were steam or vacuum distillation, solvent extraction, and heated headspace sampling. In 1977, a novel approach to the collection of trace levels of accelerants in arson debris was proposed by Twibell and Home[7] of the Home Office Forensic Science Laboratory in the United Kingdom. This approach involves placing a wire coated with carbon inside the container of debris. Vapors from the accelerant are adsorbed onto the surface of the charcoal and retained; because no sweeping of the vapors onto the charcoal is involved, this approach is known as "static" sampling. After several hours, the wire is removed and heated to desorb the vapors for instrumental analysis.

During the 1970s, acceptance of the role that laboratories could play in fire investigations led to a rapid growth in requests for examination of fire debris samples. With this increase came a need for quicker sample processing, and especially for a sampling technique suitable for both light and heavy products in a single collection. The simplicity of the carbon wire technique facilitates examination of large numbers of samples with minimal sample preparation by the analyst and is relatively effective with heavy as well as light flammable liquids.

The concept of carbon adsorption, particularly on thin layers, led to two new techniques now undergoing evaluation. In one of these, a flat carbon strip is placed in the evidence container in the same manner as the coated wire. The flat strip has much greater surface area than the carbon-coated wire and is more effective in adsorbing vapors in the container. This improved adsorption is particularly useful with samples containing low or trace levels of the accelerant. A second collection and concentration technique uses carbon granules packaged in a tea-bag configuration and hung in the container with a string.

The major differences between these and the original wire approach are the amount of charcoal used and how the sample is recovered for analysis. The amount of charcoal on the wire is very small and the charcoal's ability to adsorb liquid is thus limited. With the carbon wire, the sample is heated to desorb the accelerant; in the newer techniques, the carbon is washed with a solvent to remove the adsorbed hydrocarbons for analysis. Although the newer techniques are still undergoing development, use of the carbon strip has provided highly satisfactory results.

Variations on Purge-and-Trap Techniques

At about the same time the charcoal-wire technique was being pioneered in Europe, a new purge-and-trap technique was being evaluated in the United States.[8] In this technique, a chromatography adsorbent such as Florisil is placed in a tube attached to the evidence container and retains vapors of liquids such as gasoline and kerosene; a vacuum pump pulls room air through vent holes in the evidence container and then through the Florisil. Because of the sweeping effect of airflow through the container, this type of sample collection is known as "dynamic" sampling. On completion of sampling, adsorbed hydrocarbons are removed from the trapping medium by rinsing with carbon disulfide, an excellent solvent for petroleum hydrocarbons but, unfortunately, a toxic and unpleasant substance to use. Although effective with kerosene or fuel oil, the more volatile components of gasoline are not well retained on Florisil.

Inexpensive granular charcoal has long been used in filters, such as gas masks, to trap volatile liquid vapors. By the 1970s, purge-and-trap sample collection on charcoal granules was well established in environmental studies but had not been widely used in the forensic laboratory. However, a technique for the collection of dynamite vapors from bombing scene debris developed by Chrostowski et al.[9] was modified by them for use with arson evidence. This purge-and-trap approach uses a small glass Pasteur pipette containing a few grams of charcoal granules as the trap. One end of the trap tube is inserted into a hole in the lid of a metal paint can containing the evidence and the other connected to a vacuum line. The can is heated from the bottom during sample collection to promote vaporization of residual liquid accelerants in the evidence and to improve sampling efficiency. Air flows into the can through holes punched in its bottom, thus sweeping or purging any accelerant vapors onto the trap.[10]

With minor modifications, this purge-and-trap technique is now the generally accepted method in the United States and Canada for sampling fire debris evidence. One modification has been elimination of the vent holes in the base of the can. Instead, three holes are punched in the can lid: one for the charcoal sample collection tube, another for an inlet tube for air, and a third for a thermometer to monitor the temperature inside the can. The inlet tube is packed with charcoal to remove hydrocarbons in the laboratory room air that could otherwise enter the can, be swept onto the trap tube, and later be presumed to have originated from the debris, contributing to erroneous results. A second modification is the use of a large laboratory heating mantle enclosing the can to ensure even heating and lessen potential failure to collect flammable liquids that condense on the cooler upper surfaces of the can. This purge-and-trap system is shown in Figure 1.

The Gas–Liquid Chromatograph Arrives

Early Use of the GLC

During the late 1950s, a new analytical instrument, the gas–liquid chromatograph (GLC, sometimes called the gas chromatograph or GC), was developed. This instrument uses a long thin tube (or column) packed with a granular support material onto which is coated a thin layer of a viscous liquid such as a silicone oil or grease, called *liquid phase*. The analytical sample is injected into a heated zone ahead of the column and vaporized. A *carrier gas* flowing through the instrument sweeps the vapor onto the GLC column, which is inside an oven whose temperature is carefully controlled. As the vapor moves through the column, components of the sample dissolve, depending on their solubility in the viscous liquid. There are slight differences in solubility of individual compounds of a mixture such as gasoline; some are retained to a greater extent than others. Because the column is heated, the retained compounds are gradually released from the liquid back into the vapor stream. They are swept by the flowing carrier gas down the column where they redissolve.

This cycle from liquid phase to gas carrier is repeated again and again, ultimately resulting in a highly efficient separation of the mixture's components. Those least soluble in the stationary liquid coating move quickly through the column; those with greater solubility are retarded and exit the column later. As the separated vapor components complete their trip through the column, they pass through a detector and produce an electronic signal that rises as a component arrives, reaches a maximum, then falls off as the component completes its passage through the detector. For a given type of column and temperature, the time required for a compound to arrive at the detector varies with the chemical composition of the compound and is known as its "retention time". The intensity of the detector

Figure 1. Apparatus for purge-and-trap sample collection.

signal is a measure of the quantity of the material present in the sample. The detector signals are recorded on a moving chart to produce a pattern of peaks and valleys called a *chromatogram*.

Even first-generation GLC instruments gave, for the time, impressive performance in the analysis of petroleum products, with component separation that was orders of magnitude better than the most efficient distillations. Design limitations, however, restricted their effectiveness for the detection of accelerants in fire debris. For instance, in chromatography, separation efficiency is related to column diameter, length, and stationary phase type. Short, large-diameter columns with high concentrations of liquid phase restrict component separation and thus the resolution of one signal peak from another. Early GLC columns were typically 1/4 inch in diameter, large compared to current columns, and relatively short in length.

Another limitation is that early detectors measured thermal conductivity and reacted to a wide range of compounds, including water. Fire debris is nearly always wet, however, and a large amount of water in a sample seriously affects detection of minor components, such as traces of residual flammable liquids. In addition, because the columns were operated at a constant temperature, compounds boiling at higher temperatures were eluted slowly, resulting in long analysis times and broad peaks on the chromatogram that often overlapped, making pattern interpretation difficult. In the late 1950s, it was not unusual for several hours to be required for all the components in a complex mixture to pass through the column, and broad peaks, often lasting several minutes, were common.

Next Generation of GLC Instruments

Improvements in GLC instrumentation came quickly, however, and despite initial limitations the potential forensic applications of gas chromatography were quickly recognized. In April 1960, Lucas in Canada described the application of GLC to the identification of petroleum products in the forensic laboratory and discussed potential limitations in the brand identification of gasoline.[11] In July of the same year, other investigators, using dual-column instruments with 6-foot columns operated isothermally and using different types of liquid phases, showed that there are differences between gasolines and expressed optimism about possible brand identification.[12] Among the pioneers in the application of GLC to fire debris examination were Paul Kirk and his students at the University of California. They examined the effects of partial evaporation, or *weathering*, in the identification of accelerants in evidence collected at the scene of a fire.[13]

Regardless of the analytical technique used, products produced by thermal degradation of materials such as wood and plastics at the scene (pyrolysis products) complicate the analysis and interpretation of the results. These products and their effects were studied by Bruce Ettling[14] and later by Ettling and Adams.[15] In early work, they used a short (2-foot) column in the gas chromatograph but with a much improved detector, the flame ionization detector (FID). The advantage of FID for fire debris analysis is its high sensitivity to the hydrocarbons common in petroleum products and its insensitivity to water. Thus, with FID, trace levels of a flammable liquid are readily detectable even in the presence of the significant quantities of water usually present in fire debris.

Subsequently, Ettling and Adams employed a constant rate of heating, known as temperature programming, of the GLC column. By steadily raising the column temperature, components with higher boiling ranges are eluted more quickly, decreasing overall analysis time. Faster movement through the column lessens the tendency of a component to spread out over a portion of the column's length and limits it to a short zone. As a result, the detector signal rises and falls more quickly, making peaks on the chromatogram much sharper and with less overlap. Sharper and better-resolved peaks facilitate identification of the pattern recorded on the chromatogram as representative of a particular type of product.

Case Applications of the GLC

In late 1970, one small federal laboratory began to perform arson evidence analyses for state and local law enforcement agencies. In one of the first such cases submitted to the laboratory, the chromatograms of samples collected at the scene of a grease fire at a local restaurant bore an uncanny resemblance to gasoline. When this evidence was presented at trial, the restaurant owner was convicted of arson entirely on circumstantial evidence. This is one example of arson for profit: the owner had recently opened a new restaurant in the suburbs and attempted to "sell the old one to the insurance company", as the saying goes. Fortunately, the crime laboratory intervened in the "fire sale". The federal laboratory used state-of-the-art, dual-column GLCs with temperature programming, FID detectors, and 20-foot, 1/8-inch diameter columns. These, combined with use of a 3% liquid phase, much lower than those then widely used in arson analysis, provided improved resolution and more definitive chromatograms.[6]

The benefits of sharper peaks and more easily compared chromatograms were evident in a suspected arson in Montana. In this case, prior to the fire the defendant had gone to considerable lengths to establish that his oil furnace was leaking. Following a fire that destroyed his newly built and well-insured house, he contended that the faulty furnace was the cause. At the fire scene, the arson investigator collected oil-soaked rags found near the furnace and oily dirt from the cellar floor. He submitted these, together with a sample from the tank supplying the furnace, to the laboratory for examination. Oil extracted from the rags and the soil sample was analyzed by gas chromatography and readily identified as no. 2 fuel oil. Examination of the sample from the tank identified it as no. 1 fuel oil, or kerosene. These findings demonstrated that leaking oil from the furnace could not have been responsible for the fire and supported a contention by the state that the suspect had brought a different type of oil into the house to initiate and accelerate a fire when falling evening temperatures turned on the furnace. Although he had a well-planned alibi with proof that he was out of town when the fire occurred, the defendant was convicted of arson.

Further Improvements to the GLC

Despite the good resolution of the improved chromatographic system, in some instances it was difficult to confidently distinguish between highly similar gasoline samples. A notable instance was a request by a local agency to compare groundwater samples contaminated with gasoline with samples from storage tanks in the area. It was hoped that comparison would identify the leaking tank from among about a dozen candidates collected at service stations nearby. Although most of the samples submitted could be discriminated by GLC on silicone columns, for a few the observed differences were minor.

To improve discrimination of gasoline samples, GLC columns with a polar liquid phase were evaluated.[16] This type of liquid phase allows straight-chain, or aliphatic, hydrocarbons to pass quickly through the column while retaining aromatic, or ring-type, hydrocarbons, such as benzenes and toluenes. With the aliphatic hydrocarbons off the column, the chromatogram of the aromatic hydrocarbons alone is much simpler, affording a more definitive comparison of the aromatics in two samples. Thaman[17] had earlier combined results from polar and conventional silicone columns to enhance discrimination of gasoline samples. Also, polar liquid phases for arson evidence examination are receiving new attention because the requirement for automotive use of unleaded gasolines, containing much higher levels of aromatics than leaded gasolines, has made them widely encountered in arson cases.

As previously indicated, decreased diameter and increased length of the GLC column provides improved peak resolution and more comparative information. The first reported application in a forensic laboratory of such a column was by Cain in the United Kingdom, who obtained excellent results by using a very small diameter, or capillary, glass column 60 meters in length for comparing kerosene samples.[18] In addition to its reduced diameter, the capillary column was an open tube with a thin layer of liquid phase coating the inside of the tube wall. Coating the liquid on the wall, rather than on a granular support packed into the column, reduces problems with high back pressure and low flow rates that become serious as the length of a packed column is increased beyond about 20 feet. The open tube configuration allows column length to be increased virtually without limit and gives much-improved peak separation and sharpness. In the United States, Armstrong and Wittkower pioneered capillary column GLC for the examination of fire debris and examined a number of the products of pyrolysis during fires.[19]

Although capillary columns had been available for a number of years and were evaluated by Wineman in the mid-1970s, several factors restricted their acceptance in forensic laboratories. First, a basic patent on the technology limited availability to a single supplier, making them costly for routine use. Furthermore, metal capillary columns, although sturdy, are not chemically inert and tend to promote on-column decomposition of certain types of compounds, particularly at elevated temperatures; glass columns can be made chemically inert, but because of their fragility they are extremely difficult to install or remove without breaking.

These limitations were overcome with the introduction of capillary columns made of fused silica, with inside diameters not much larger than that of a human hair. They are tough and chemically inert, yet they are extremely flexible so column breakage is no longer a significant problem. One limitation of early fused silica columns was their inability to operate at the extremely high temperatures required for analysis of heavy oils or candle waxes, which are encountered occasionally in arson examinations. This limitation was overcome with a thin metal coating on the outside surface of the column, extending its operating range by several hundred degrees. Fused silica capillary columns are now the accepted standard for arson examinations and, even in relatively short lengths, provide far greater resolution and better peak shapes than packed columns. For comparison, gasoline chromatograms from packed and capillary columns are shown in Figure 2. The development of capillary columns, with their ability to separate individual compounds in complex mixtures, made the next significant development in arson evidence examinations possible.

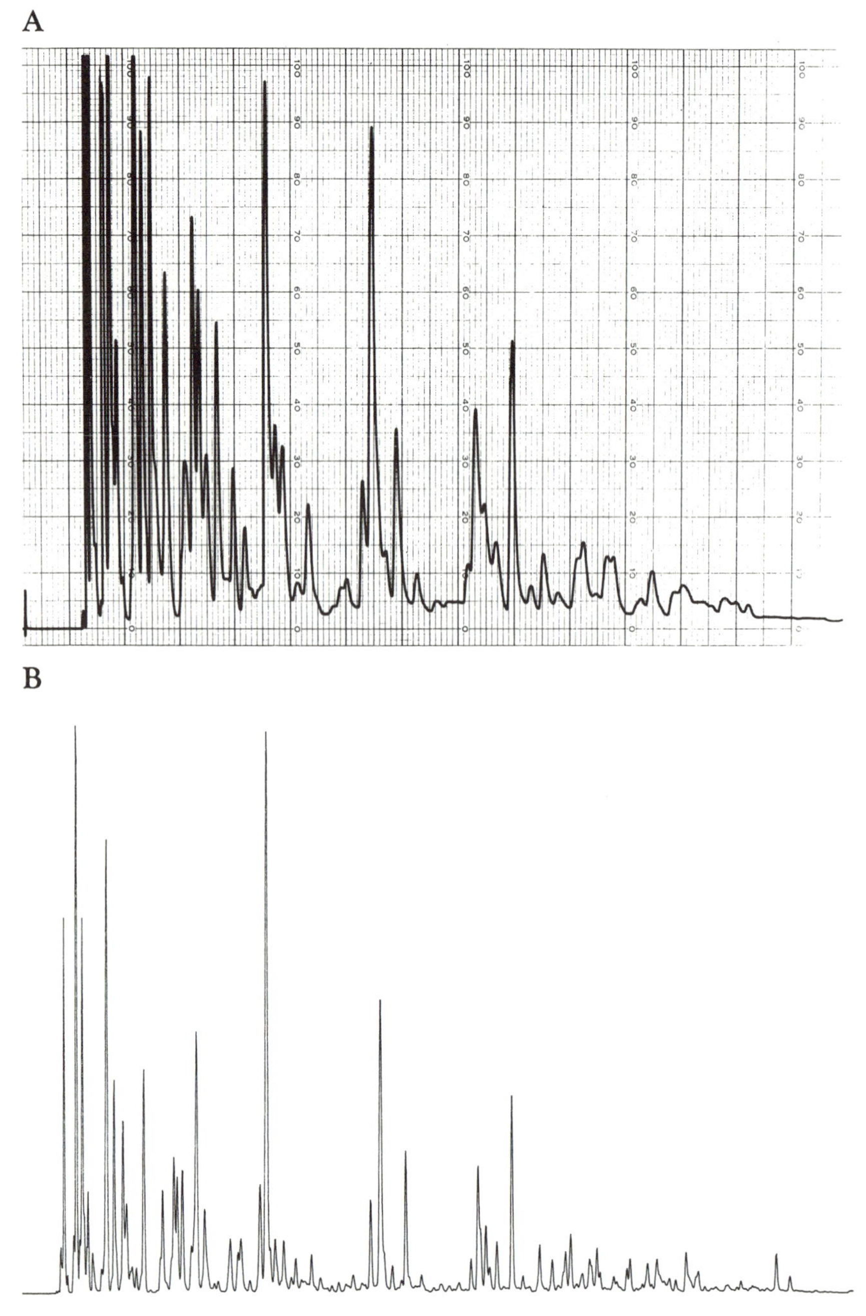

Figure 2. Gas-chromatograph analysis of unleaded gasoline. A. Packed column. B. Capillary column.

Mass Spectrometry Increases Precision

As previously discussed, a major complication in arson evidence analysis is the production of pyrolysis products during the fire. These contribute to the material separated from fire debris and can make the analytical sample extremely complex. Some of the chemical compounds produced during pyrolysis are among those found in common flammable liquids, posing additional problems in interpretation of the chromatogram. Because of the masking effect of pyrolysis products, it may be difficult to determine whether an accelerant is present in the sample. In addition, even in the absence of pyrolysis products, as a particular product, such as gasoline, is steadily evaporated, its composition shifts toward increasing percentages of the components with a higher boiling range. As a result, the chromatogram of a highly weathered gasoline tends to look more and more like that of kerosene or fuel oil. In general, the analyst has no knowledge of the history of a fire debris sample submitted for laboratory examination. In a few instances when only the chromatographic pattern of a highly weathered sample is relied on for identification, the accelerant might be mistaken as being a midrange distillate when it is actually gasoline that was extensively evaporated during the fire.

The mass spectrometer (MS), an instrument that measures the mass of intact molecules and fragments produced by breakup of the parent molecule, can help overcome these problems. A number of techniques are available to produce fragments or ions from a chemical compound, but under controlled conditions fragmentation is reproducible and the fragments and their relative ratios serve to identify the original compound. With its ability to identify a particular species, something not possible with conventional gas chromatographic detectors, the mass spectrometer provides a new dimension in detection for the gas chromatograph. The gas chromatograph–mass spectrometer (GC–MS), which combines the ability of the capillary column gas chromatograph to effectively separate complex mixtures into individual components with the ability of the mass spectrometer to uniquely identify each of the separated components, is an especially powerful tool for use with arson evidence.

The identification capabilities of the GC–MS combination greatly enhance examination of samples highly contaminated with background materials, as well as verification of the identity of highly evaporated accelerants. In 1976, Zoro and Hadley in the United Kingdom evaluated forensic applications of GC–MS, including its use in the examination of headspace samples from fire debris.[20] In a sample containing pyrolysis products, they identified several compounds typical of flammable liquid accelerants by monitoring the mass spectrometric data for specific ions. One year later, Mach at Aerospace Corporation studied simulated arson residues by using GC–MS to identify compounds typically present in gasoline that could be used as indicators.[21] In his study, a series of samples of gasoline were evaporated to increasingly higher degrees to simulate residual material found at fire scenes.

The following year, the use of GC–MS to examine actual fire debris samples was reported.[22] In this work, both headspace and steam-distilled samples were examined. The value of the approach was illustrated by showing that GLC peaks in certain suspect samples were due to terpenes, naturally present in pine and the major ingredients of turpentine. In debris from a fire involving coniferous wood, particularly when steam distillation is used for sample recovery, these should be expected as background contaminants. Unless encountered in extremely high levels and primarily as diterpenes, they do not indicate turpentine as an accelerant, although their use for that purpose cannot be precluded. Several aromatic compounds, when found in samples, are considered indicative of the use of gasoline as an accelerant, whereas esters and ketones suggest the use of lacquer thinners or similar mixtures.

The availability of computer data systems capable of efficiently searching a file of all the MS data collected during an entire gas chromatographic "run" of an arson sample, rather than looking for only specific ions, expanded the horizons of GC–MS. With this approach, which he called *mass chromatography*, Smith[23] looked for peaks characteristic of particular accelerants and applied the technique to actual case samples. Some of the case samples contained charred debris from carpet and carpet backing that contributed peaks to the chromatogram. The result was a pattern whose interpretation was equivocal with GLC analysis alone. Using the mass chromatography approach, several of these were shown to contain residual accelerants; in others no accelerant was found. In one case, fuel oil was identified on the shoes of a suspected arsonist.

Although a system for classification of petroleum-based accelerants into five general categories had been widely accepted, it was recognized that many of the materials detected in actual cases did not fit neatly into any one. Classification of a product into a particular category was made by observation of GLC patterns to determine boiling range, but the apparent boiling range of a suspected accelerant could be distorted by contributions from pyrolysis products or loss of a significant fraction of the original accelerant through evaporation. With its selectivity as a detector for GLC, MS could be of significant value in improving product classification, particularly in the presence of pyrolysis products.[24] In 1990, Nowicki[25] used GC–MS data, in addition to that from the chromatogram, to develop a more detailed accelerant classification system. In addition to the GC–MS identification of ions typical of known accelerants, Nowicki emphasized the chemical nature of the components present in the sample. He proposed a system for classification encompassing eight categories. Light and heavy products were divided into two categories each, and a new category was devised to include synthetic products that are rigorously not petroleum products. The combination of boiling range information from the GLC with GC–MS data on actual chemical composition has potential for improving identification of highly weathered or "fire-aged" products and lessens some interpretation problems caused by the presence of pyrolysis products in the sample.

Another way of resolving complex chromatograms from highly contaminated samples using GC–MS was taken by Wineman with a "target compound ratio" approach.[26] In this work, compounds typical of particular classes of accelerants but uncommon in pyrolysis products, were identified and their mass spectra cataloged in a computer-searchable reference library. The ratios of pairs of reference or "target" compounds with boiling points close together were determined. The premise was that when evaporation of the accelerant occurs during a fire, closely boiling compounds should evaporate at similar rates and the ratios of these compounds should remain essentially unchanged even with extensive evaporation. In the presence of high levels of background contamination, the target compound ratio may be used with high confidence for flammable liquid identification. Following the GC–MS analysis of the sample, a computer search for the target compounds is made; if any are found, ratios of the selected target compounds in the scene sample are compared with those observed for the same compounds in known classes of accelerants. To test the effectiveness of this method, samples containing high levels of pyrolysis products were spiked with evaporated accelerants to produce "worst-case scenario" samples. In these tests, good detection and identification of the added accelerant were obtained in the presence of pyrolysis products from a variety of materials common at fire scenes. Since Wineman's retirement, development of this technique has been continued by Keto.[27] It is used as needed to confirm or eliminate the identification of an accelerant in high-background samples examined by capillary column GLC.

Techniques for the Future

Currently, adsorption of vapors on charcoal, analysis of the extract from the charcoal by capillary column gas chromatography, and, for particularly difficult or highly contaminated samples, confirmation by GC–MS is state-of-the-art science for examining fire debris evidence. But forensic science, like other science, is not static. New products appear on the market and soon turn up as evidentiary materials in criminal cases. New instruments and ideas for using established techniques from other technical fields provide potential new tools for the crime laboratory. Some of these we are just learning to use; others are ready to be taken out of the forensic toolbox and put to work. Each can assist in improving our capabilities for the task of arson evidence examinations.

In addition to further development of mass spectrometric techniques to examine contaminated samples, other techniques offer potential for the crime laboratory to provide more effective support in the investigation of arson. Examples include microwave heating for debris samples and desorption of accelerants captured on charcoal. This approach offers simplicity, speed, and elimination of the need for hazardous solvents, improving safety in the laboratory and benefiting the environment.

For recovery of accelerants from fire scene debris, supercritical fluid extraction (SFE) is a technique of considerable potential. SFE uses carbon dioxide, maintained under pressure as a liquid, as an extraction solvent. Supercritical fluids offer several advantages over extraction by a flammable or toxic liquid solvent and over collection and concentration of vapor. Because the CO_2 used for the extraction is a liquid, it has good penetration of the sample and recovery of compounds covering a range of molecular weights typical of liquid extractions, yet it is nonhazardous. By control of the extraction conditions and the addition of small amounts of modifiers, the extraction characteristics of the supercritical fluid can be changed, perhaps lessening the extraction of background contaminants from the debris. CO_2 is a gas at normal pressure, so once the sample extraction is complete, simply reducing the pressure removes the solvent and leaves the extracted material essentially intact for analysis. This gives the analyst a range of options for analysis, depending on the information required, and greatly facilitates interpretation of the results.

Other approaches of promise include capillary column gas or supercritical fluid chromatography with two or more different types of detectors. Multiple detectors permit, in a single analysis, simultaneous or sequential detection of polar or oxygenated species, such as alcohols or methyl tertiary butyl ether (MTBE) added to unleaded gasoline to improve engine performance. Detection of these in addition to the expected hydrocarbons increases confidence that a sample contains a residual accelerant such as gasoline and not a product solely of materials normally present at the scene. Alternatively, parallel or sequential GC columns of different types can improve chromatogram resolution, shorten analysis time, and simplify results. With a dual-column system, a single injection of a sample can simultaneously and separately be examined for aliphatic and aromatic hydrocarbons, hydrocarbons and alcohols, or other combinations. Splitting the sample puts each half on a column, providing the best resolution for a particular class of compound. The results from both columns are recorded and displayed, providing twice the information normally available in a single run. Similar results can be obtained by connecting columns with two different types of liquid phases end to end. Besides reducing analysis time, the additional identification provided by dual columns assists the examiner in identifying samples meriting further examination. It may also assist with samples whose chromatograms, although complex, are readily attributable to typical pyrolysis products, thus eliminating additional unproductive analyses.

Another approach, well suited to the examination of heavier products such as lubricating oils, is high-performance liquid chromatography (HPLC). Because it operates at room temperature, HPLC does not thermally degrade high-boiling-range products as GLC does, and combining it with the mass spectrometer offers the same advantages as GLC. Interfacing HPLC with the mass spectrometer has been difficult, but newer instruments are rapidly overcoming the problem.

As time passes, lower-cost mass spectrometers and development of MS detectors designed specifically for use with gas chromatographs are ensuring wider use of GC–MS in arson analysis. Also, the rapid increase in power and falling cost of computers will lead to greater use of automated sample analysis and data collection. Expert systems—computer systems capable of "learning"—will further lessen demands on the analyst in the examination of complex chromatograms. With the additional information they provide, these newer approaches offer the potential for improved reliability and efficiency in the laboratory.

The laboratory examination of fire scene evidence to aid in the detection of arson has come a long way since the work of the early pioneers. Because arson is a major crime that poses a threat to society far beyond its intended victims, combating it must have a high priority. With the justice system's need for technical help in investigating and prosecuting this crime and with demands by the courts that more use of scientific evidence be made in criminal investigation, arson analyses will continue to challenge forensic scientists. In meeting this challenge, chemists will develop new and improved laboratory techniques to improve the certainty of detection of traces of residual accelerants while lessening the chance for an accelerant to be identified in a sample when none is present. Although dedicated and motivated scientists work daily in examining arson evidence, those who develop and use innovative approaches to examine fire debris samples will be the new pioneers in the laboratory examination of arson evidence.

References

1. Bennett, G. D. *J. Crim. L. Criminol. Police Sci.* **1958,** *49*(2), 172–177.
2. Macoun, J. M. *Analyst* **1952,** *77*(916), 381.
3. Bennett, G. D. *J. Crim. L. Criminol. Police Sci.* **1954,** *44*(4), 652–660.
4. Brackett, J. W., Jr. *J. Crim. L. Criminol. Police Sci.* **1955,** *46*(4), 554–561.
5. Adams, D. L. *J. Crim. L. Criminol. Police Sci.* **1957,** *47*(5), 593–596.
6. Midkiff, C. R., Jr.; Washington, W. D. *J. Assoc. Off. Anal. Chem.* **1972,** *55*(4), 840–845.
7. Twibell, J. D.; Home, J. M. *Nature (London)* **1977,** *268*(5622), 711–713.
8. Baldwin, R. E. *Arson Anal. Newsl.* **1977,** *1*(6), 9–12.
9. Chrostowski, J. E.; Holmes, R. N.; Rehn, B. W. *J. Forensic Sci.* **1976,** *21*(3), 611–615.
10. Chrostowski, J. E.; Holmes, R. N. *Arson Anal. Newsl.* **1979,** *3*(5), 1–17.
11. Lucas, D. M. *J. Forensic Sci.* **1960,** *5*(2), 236–247.
12. Cadman, W. J.; Johns, T. *J. Forensic Sci.* **1969,** *5*(3), 369–385.
13. Parker, B. P.; Rajeswaran, P.; Kirk, P. L. *Microchem. J.* **1962,** *VI*, 31–36.
14. Ettling, B. V. *J. Forensic Sci.* **1963,** *8*(2), 261–267.
15. Ettling, B. V.; Adams, M. A. *J. Forensic Sci.* **1968,** *13*(1), 176–89.
16. Midkiff, C. R., Jr. *Arson Anal. Newsl.* **1980,** *3*(6), 1–22.
17. Thaman, R. N. *Arson Anal. Newsl.* **1976,** *1*(1), 11–19.
18. Cain, P. M. *J. Forensic Sci. Soc.* **1975,** *15*(4), 301–308.
19. Armstrong, A. T.; Wittkower, R. S. *J. Forensic Sci.* **1978,** *23*(4), 662–671.
20. Zoro, J. A.; Hadley, K. *J. Forensic Sci. Soc.* **1976,** *16*(2), 103–114.
21. Mach, M. H. *J. Forensic Sci.* **1977,** *22*(2), 348–357.
22. Stone, I. C.; Lomonte, J. N.; Fletcher, L. A.; Lowry, W. T. *J. Forensic Sci.* **1978,** *23*(1), 78–83.
23. Smith, R. M. *Anal. Chem.* **1982,** *54*(13), 1399A–1409A.
24. Bertsch, W.; Sellers, C. S. *LC-GC* **1988,** *6*(11), 1000–1014.
25. Nowicki, J. *J. Forensic Sci.* **1990,** *35*(5), 1064–1086.
26. Wineman, P. L. "A 'Target Compound Ratio' Approach to the Detection of Petroleum Fuels and Solvents in Highly Contaminated Extracts of Fire Debris"; unpublished report; Bureau of Alcohol, Tobacco, and Firearms: Rockville, MD, 1985.
27. Keto, R. O.; Wineman, P. L. *Anal. Chem.* **1991,** *63*(18), 1964–1971.

6

The Microscope in Forensic Science

Forensic Microscopy in the 1890s and the Development of the Comparison Microscope

David A. Stoney and Paul M. Dougherty

In his history of criminalistics and forensic science education in the United States, Dillon identifies the period from 1880 to 1900 as the beginning of criminalistics practice, when "the first significant American efforts were initiated by microscopists".[1] He goes on to point out that the extensive literature developed at this time was essentially lost and that the "efforts of these scientists terminated around 1900 because of factors including the nature of the American police".[1] Dillon's comments inspired us to look further at this period in forensic science practice.

Two primary sources of information were used in this research: the microscopy literature appearing in the journals of the day and forensic medicine textbooks. Hans Gross's *Criminal Investigation*,[2] which is often taken as a starting point for early criminalistics cases, was published after 1900 and so falls outside the years of interest here. Cases and forensic science practice summarized by Gross that refer to earlier work have been included.

In this chapter, we undertake four things. First, we review literature that was available to a forensic microscopist in the 1880s and 1890s. What reference books could have been on the microscopist's shelf and to what extent did they treat materials of interest to forensic scientists? Second, we address the extent to which forensic science philosophy or practice was developed. Were people looking at evidence the same way we do today? What perspective did they have, and was the "spirit" of criminalistics present? Third, we examine specific types of trace evidence in detail by looking at cases prior to 1880 and between 1880 and 1900. Although microscopy was used considerably in toxicology and serology during this period, we just touch on these better-known applications and emphasize cases with evidence in the form of hairs, fibers, soil, and miscellaneous types of stains. Fourth, we look at an important forensic laboratory tool, the comparison microscope, from its roots in the 1880s to its refinement in the four decades that followed.

Literature Available to the Forensic Microscopist in the 1890s

On the forensic microscopist's bookshelf in the 1890s were a number of impressive atlases and treatises describing materials of interest in the forensic sciences and specifically addressing many forensic science problems. These may be generally classified into works on microscopy, works on forensic medicine, and works on food and drug analysis.

One of those works that is still of exceptional scientific value is Griffith and Henfrey's *Micrographic Dictionary.*[3] Volume 1 is an encyclopedia with descriptions and citations of works on all kinds of microscopic objects. Volume 2 consists of plates that are referenced in Volume 1: there are 75 plates with hundreds of illustrations that include biological materials, common household substances, and drugs. Particularly relevant plates are those showing diatoms, foraminifera, polarizing objects, pollen, rocks, starch, and hairs. Another impressive book on the shelf would be Hassall's text[4] on the adulteration of foods, which contains detailed discussion and illustration of food microscopy including sugar, honey, flour, and starch. During the 1880s and 1890s, crises were developing that preceded the passage of the first Federal Food and Drug Act in 1906.[5] Hassall exhaustively considers adulterants and poisons that may find their way into foods. At a time when poisons were used for coloring in icings and cake decorations, Hassall shows a well-developed sense of the importance and necessary requirements of forensic science practice. He also has an exceptional understanding of the importance of microscopy in analytical laboratory work:

> So great and manifest are the differences revealed by the microscope in various vegetable substances, that, with ordinary care and some amount of preliminary knowledge, the discrimination becomes a matter of the greatest ease and the most absolute certainty....
>
> Further, wonderful to relate, the grinding and pulverization, and even the charring, of many vegetable substances, does not so destroy their structure as to render their identification by the microscope impossible....
>
> Again, substances may be discovered by means of the microscope, even when introduced into articles for the purpose of adulteration in extremely minute quantities....
>
> There is still another use to which the microscope may be applied...; it may frequently be made to serve as an auxiliary to chemical researches: thus, for example, when we want to ascertain whether any substance contains starch, carbonates, phosphates, etc., it is often the quickest and most certain way to apply the reagents to a small quantity of the substance while this is under the field of vision of the microscope. Chemistry can tell us whether starch is present in any substance, but it is very seldom indeed that it can furnish us, as the microscope so constantly does, with the name of the plant from which the starch was derived: it can indeed also make us acquainted with the fact that woody fibre is contained in any particular article, but it cannot furnish us with the name of the tree or plant of which it formed a constituent. Another great advantage of the microscope over chemistry is the greater speed with which results may be arrived at. Many chemical analyses occupy days, while most microscopical examinations may be made by the practised observer in the course of a few minutes....

Moving into toxicology, we would have access to Wormley's treatise on the microchemistry of poisons[6] with his wife's exceptional engravings of microcrystalline products obtained through sublimation or microchemical reactions. We would also have Blyth's work[7] on poisons, which includes characterization of crystalline products obtained by recrystallization from water, sublimation, electrodeposition, and microchemical reactions. Taylor's *Medical Jurisprudence*[8] as well as Tidy's *Legal Medicine*[9] and Witthaus and

Becker's *Medical Jurisprudence*[10] are more general works covering a variety of forensic microscopy topics.

Four other specialized monographs serve as examples of the development of forensic science practice. We have Ballowitz's work[11] on spermatozoa, Formad's book[12] on the microscopical examination of blood stains, Moeller's atlas[13] of pharmacognosy, and van Ledden Hulsebosch's work[14] on the examination of human feces. These monographs are of a quality and thoroughness that would impress if not shame many forensic scientists today.

Based on the literature of the time, certainly there was established expertise in forensic microscopy in the 1890s. But did practitioners understand forensic science and the practice of forensic microscopy and trace evidence analysis the way we do today?

General Practice of Forensic Science in the 1890s

It is satisfying and inspiring to discover in the early works on forensic medicine that the description of our specialty is given with both the eloquence of the era and with an understanding and fervor that reminds us of our privilege to practice in the profession and the attending responsibility. Tidy's *Legal Medicine* contains one of the best descriptions of the challenge that defines our work:

> To shirk the manifest conclusions of experiment, because such conclusions are not in harmony with the generally received opinions of authorities, is unscientific timidity. On the other hand, dogmatically to state as facts the conclusions drawn from one's own experiments... may seem to partake of scientific venturesomeness. And those only who in Science work have felt the difficulty, and yet the necessity, of avoiding the extremes of timidity and venturesomeness, can realize how that difficulty becomes intensified when, as in Legal Medicine, the subject relates to such serious issues as imprisonment or freedom, and it may be life or death.
>
> For the medical jurist, whose object should be the interests of justice, to hesitate where science is positive, is as unjustifiable as for him to speak without reserve on those details of our science, where the limits of exact scientific knowledge are undefined. There is a scientific certainty which only the coward treats as uncertainty, and there is an uncertainty which only the boldness of ignorance ignores.[9]

Microscopy and Trace Evidence in the 1890s

Microscopy Applied to Toxicology and Serology in the 1890s

Two applications of forensic microscopy, toxicology and serology, are dealt with superficially here. This is done to avoid overemphasizing them and because they were well developed by the 1890s insofar as they were practiced at that time. Earlier reference was made to Wormley's and Blyth's work[6,7] in toxicology. Taylor's *Medical Jurisprudence*[8] serves as another example. Taylor made extensive use of microscopy in the systematic analysis of poisons that includes 43 drawings of crystals, each with magnifications given. Each shows variation in habit, and each has a text discussion that is sensitive to the habit modifications that might occur.

In forensic serology, the microscopical examination of blood and semen stains was performed with a level of detail and interest motivated by the absence of any immunological or enzymatic tests. Taylor's *Medical Jurisprudence*[8] gives a particularly detailed protocol for bloodstain examination. Stains, depending on their quality, could allow examination of the red blood cells themselves. One specific issue of extreme importance was the degree to which the micrometry of these reconstituted blood cells could be used to infer the species of the blood. Formad's work[12] as well as the medicolegal works cited[8,9,15] document the careful and critical attention given to this area. There was considerable professional concern of the possibility of overinterpretation of the results.

But these early attempts at determining the possible species of origin of blood were only part of the microscopical analysis of a bloodstain. The criteria for a positive bloodstain identification included seeing the remains of the cells, a positive crystal test for heme, and an examination of the stain extract with a microspectroscope to observe changes in the absorption spectrum resulting from microchemical testing.

Identification of semen stains began by looking at the character of the stain. Presumptive observations included feel, color, and responses to heat (yellowing), to moisture and heat (odor), and to nitric acid (no precipitation and a color change to yellow). These presumptive chemical tests were considered worthless if there was dirty fabric, and in such cases one could rely only on microscopical examination. In any event, no conclusive opinion would be offered without positive microscopical identification of spermatozoa.[9,15] There was considerable discussion of alternative methods for extraction of semen stains and staining of the cellular material. The need for intact spermatozoa for positive identification was, as it is today, disputed by some, but Tidy cautioned that there are many look-alikes for detached sperm heads. He felt that if the spermatozoa were not intact, "one should admit nothing".[9]

In addition to the identification of a semen stain, particular care was taken to observe any other trace evidence that might be present. Thus, Wood states:

> Everything associated with the spermatozoa should be carefully noted, since it may happen that some substance may be found adherent to the seminal stain which will tend to connect the guilty person with the crime in a case of rape. For instance, we may detect fibres of cloth of various kinds, or fragments of hair, which by some peculiarity of material or color may be identified as belonging to the accused; other substances such as pus and blood corpuscles, or cellular elements of various kinds, may also be recognized. Such extraneous substances should always be carefully noted whether they appear at the time to have any importance or not.[15]

Hair Evidence

The potential of hair for use as trace evidence was well appreciated in the 1890s. The texts at the time emphasized that any hair found on the hands or about the body should be preserved, mounted, and given careful microscopical examination. Specific places to look for hairs included the crime scene near the remains of a body, in hands, on clothing, on the accused, on the victim, on weapons, and mixed with semen or bloodstains. In particular, Park noted that bloodstains found on weapons "should be examined with the microscope, since from the detection and identification of hair or fibres of fabric evidence of the greatest value may be adduced."[16] Tidy concurred that "all blood spots… should… be examined with a large hand magnifying lens for hair, fibres, etc."[9]

Searching for the two-way transfer of hairs in sexual assault cases and in bestiality cases also was practiced. Citing cases dating from 1851 through 1875, Tidy stated that the microscopical examination and identification of hairs and fibers many times had led to the conviction of criminals.[9]

As for the technical aspect of hair examination, Émile Pfaff's work[17] (1866) clearly documented the major structural elements of hair, including pigment granules, cuticle variations, hair variations seen among different body areas, and ranges in diameters. Racial and sexual trends in the appearance of hairs also were appreciated. As is the practice today, lower magnifications were considered better for examination, and polarized light microscopy in particular was seen as having an advantage.[9] Tidy advocated direct comparison of hairs (as opposed to description) for evaluating the possibility of common origin: "The comparison of hair with hair, or of fibre with fibre, as to form, color, length, breadth, etc., yields at all times more valuable evidence, and furnishes data for more exact conclusions, than mere detailed information respecting the microscopical characters of an individual hair or fibre".[9] Tidy went on to emphasize that this advantage of direct comparison meant that one should compare hairs of unknown origin directly with standards kept in the laboratory. (There is no indication, however, that any type of comparison microscope was employed in this process. See the section on the development of the comparison microscope at the end of this chapter.)

For characterization of animal hairs, a mounted reference collection was considered essential. The distinction between human and animal hairs was recognized as usually easy to make, with some exceptions. Animal hairs were characterized by medullary index, shape, microscopical appearance, pigment location, and medullary type.[9] Animal hairs were considered particularly relevant in bestiality cases. Rosse stated:

> Unless taken in the act, the only medical evidence of value in cases of bestiality is the identification of *hairs* of the animal on the accused, or finding of suspicious spots, stains, etc., upon the clothing. It may, however, be difficult to prove the origin of the several spots as one and the same, or that the blood corpuscles or spermatozoids discoverable by the microscope are those of man or some other animal.[18]

More impressive than technical capability is the appreciation of limitations in the interpretation of microscopical hair comparisons. Tidy stated that "the observations to be valuable should be numerous" and "in giving evidence, it would be safer to say that two hairs are *similar*, than that they are *identical*".[9] Taylor states:

> An opinion of identity based on a similarity of hairs... should be expressed with caution.... There are many persons who have hair similar in colour, size, and length; hence a witness may be able to say that there was similarity, but he can rarely be in a position to swear that there is absolute identity.[8]

Wood felt that even when a correspondence was seen it could "only be stated that the hair in question is *similar* to that of the individual mentioned and *may* have originated from that individual".[19]

Table I gives a chronological listing of some important cases involving hair evidence. (See table for specific case citations.) The earliest case listed is *Regina v. Teague* (1851).[8,9] In this case a hammer was suspected as a murder weapon. The victim's injuries had included blows to the head overlapping the eyebrows. Short, stiff, white hairs corresponding to the victim's eyebrows were found on the hammer. The hammer was found on a hedge near where the suspect was working. The suspect's story was that the hammer was being used to beat goat skins, which also were present on the hedge. This focused the issue on whether the hairs were indeed human eyebrow hairs or those from a goat. The suspect had originally maintained that the deceased received the injury from a fall or from a kick from a horse. At trial, "the witnesses were severely cross-examined upon the structural differences of the hair of man and animals".[8]

In the second case, *Regina v. Hansen* (1856), hairs on a suspected murder weapon again came into evidence.[8,9] Hair found on a bloody stone corresponded to the hair of the deceased. The stone was believed to have been the murder weapon and the accused had been seen with the stone.

In *Regina v. Steed* (1863), the victim's injuries were believed to have been caused by kicking.[8,9] Hair and red wool fibers were found under the nail heads in the suspect's boots. These matched the victim's head hair and the comforter that she was wearing. Similarly, in *Regina v. Divine* (1864), gray hairs corresponding to the deceased were found on a poker that was believed to be the murder weapon.[8,9]

The case of Rosetta Bishop (1864) was a murder–suicide.[8,9] Harley's examination is summarized by Taylor:

> Harley found on a hatchet certain hairs from one to three and one-half inches long, which he described as human hairs from the head of a fair person who was becoming grey. From their fineness he considered them to be hairs from the head of a woman, and when compared with those taken from the head of the deceased woman (*Rosetta Bishop*), they presented so great a similarity as to leave no doubt that the hair had belonged to the same person, and that the wounds on the head had been inflicted with this hatchet.[8]

In 1865, in a bestiality case, Kutter found short, dark, pointed hairs on the penis of the defendant. These corresponded to hairs from the back part of a mare.[8] In 1887, in a similar case (*Regina v. Brinkley*), Stevenson found stains on the accused's trouser flap consisting of spermatozoa and, in addition, hairs corresponding in color, form, and length to those of a mare.[8]

Table I. Case Examples of Hair Evidence.

Year	*Case and Reference*
1851	*Regina v. Teague* (8, Vol. 1, pp 528, 565; 9, Vol. 1, p 239) Hairs on suspected murder weapon matching victim's eyebrow hair.
1856	*Regina v. Hansen* (8, Vol. 1, p 561; 9, Vol. 1, p 238) Hair on suspected murder weapon matched to deceased.
1863	*Regina v. Steed* (8, Vol. 1, pp 562, 575; 9, Vol. 1, pp 238, 239) Fibers and hairs on prisoner's boots matched to victim's hair and clothing.
1864	*Regina v. Divine* (8, Vol. 1, p 565; 9, Vol. 1, p 239) Hairs on suspected murder weapon matched to deceased.
1864	Case of Rosetta Bishop (8, Vol. 1, p 565; 9, Vol. 1, p 239) Hairs on suspected murder weapon matched to deceased.
1865	Kutter's bestiality case (8, Vol. 2, pp 471, 472) Hairs on suspect's genitals matching mare.
1867	*Regina v. Watson and Wife* (8, Vol. 1, pp 565–567) Burned residue of hat identified.
1871	Eltham murder case (9, Vol. 1, pp 242–244) Human hair comparison, hair on clothing, weapon.
1875	*Regina v. Wainwright* (9, Vol. 1, pp 230, 231) Hairs on a shovel matched to deceased.
1887	*Regina v. Brinkley* (8, Vol. 2, p 471) Hairs on suspect's clothes matching mare (bestiality case).
1889	Cronin case (*20*, pp 407, 408; *21*) Hair in trunk and bleached hair on soap matched to deceased.
~1890	Stevenson's kicking case (8, Vol. 1, p 563) Suspect cleared: alleged hairs were not hair.

One of the more intricate cases reported was *Regina v. Watson and Wife* (1867).[8] In that case suspicion of foul play was raised in the death of a man named Raynor, in part because of his missing hat. Dragging marks and shoe prints led investigators to a house where Watson and his wife lived. A search was made for the hat. No hat was found, but an iron rake fouled with a peculiar, burned substance was found concealed on a shelf. The defendants claimed that the rake had last been used to clean out a cesspool. The substance on the end of the rake was submitted for examination.

When heated, it smelled of burned shellac. When treated with alcohol, it formed a resinous solution. A microscopical examination of the alcohol extract revealed animal hairs and resin. Based on the results, a hat similar to the missing one was purchased and burned. Direct comparison of these residues with those found on the rake showed they were "precisely similar". Taylor provides us with an illustration of the microscopical appearance of the rabbit hair and shellac mixture.[8]

In the Eltham murder case (1871), human hair comparisons linked clothing and a weapon with the deceased.[9] In *Regina v. Wainwright*, hair comparisons linked a shovel with the deceased.[9]

In one of Stevenson's cases, the defendant was accused of a kicking assault and gray hairs were found on his boots. Stevenson showed that these were thistledown.[8]

Finally, in the celebrated Cronin case,[20,21] hair in a trunk and bleached hair on a bar of soap were matched to the victim:

> Subsequent evidence of medical experts was conclusive as to the identity of the hair found clinging to a trunk, the hair cut from the head of the murdered man, and that of a single hair discovered on a cake of soap. This single strand, being lighter in color in some portions than in others, seemed to indicate that it could not have come from the head of the deceased, whose hair was brown. But it was shown that hair placed on soap or other alkaline substances becomes bleached in a manner similar to the color of a single thread. This evidence of vital importance linked the hair found in the trunk with that cut from Dr. Cronin's head, and went far towards proving that one of the murderers had washed his hands with the soap after the deed had been done.[20]

Fiber Evidence

Moving on to fiber evidence, we must remember that the earliest commercially important synthetic fibers (nitrocellulose and rayon) were not developed until 1890[22] and so would not figure greatly in the casework of the period. Along with hairs, fibrous trace evidence was restricted therefore to silk, wool, cotton, linen, and other vegetable fibers. As for understanding its role as associative evidence, Tidy states:

> Fibres of different kinds and colors may be found on the weapons supposed to have caused the death, or on the person of the accused, and the question may then arise whether they correspond or not

with fibres taken from articles of clothing worn by the deceased at the time of the murder.[9]

In burial cases, soil was sifted to recover fragments of clothing.[9] Ligatures in strangulation cases were to be carefully examined for "marks of blood, for adherent hair, or other substances".[23]

Microscopical methods for the examination of the natural fibers of the day were well developed. Taylor gives illustrations at 300 diameters of cotton, linen, silk, wool, ancient wool, and ancient linen, along with human hair.[8] Tidy gives a good discussion of natural fiber identification. For cotton, he cites the flattened twisted appearance and the borders that define the lumen; for linen he describes "jointed markings"; for silk, fibers free of markings and highly refractive of light; for wool, a distinct cortex, flexible and wavy; for hemp, fibers coarser than flax, with no spiral streaks when boiled with nitric acid; and for coir, bundles of small, blunted fibers that form wide-open cross marks when boiled with nitric acid.[9]

Table II summarizes some early fiber cases. (See table for specific case citations.) Of these, *Regina v. Steed* (1863) has already been discussed in the

Table II. Case Examples of Fiber Evidence.

Year	*Case and Reference*
1852	*Regina v. Harrington* (8, Vol. 1, pp 561, 562) Fibers matching victim's clothes in blood on suspected murder weapon.
1858	New Orleans rape case (8, Vol. 2, p 461) Fibers matching suspect's clothes in blood and semen stain.
1860	*Regina v. Cass* (8, Vol. 1, pp 542, 562; 9, Vol. 1, p 239) Fibers matching suspect's clothes in blood on murder weapon.
1863	*Regina v. Steed* (8, Vol. 1, pp 562, 575; 9, Vol. 1, pp 238, 239) Fibers and hair on suspect's boots matching deceased's clothing and hairs.
1875	Littlejohn's infanticide case (9, Vol. 1, pp 222–223) Thread tying newborn's umbilical cord matched to sample in accused mother's possession.
1890	Vail case (24) Singed fibers examined in shooting reconstruction to establish position of the weapon.
1898	*Regina v. Kerr* (8, Vol. 2, pp 467, 468) Fibers matching victim's clothing found on suspect's clothes, in mud stains matching crime scene.

hair section. In *Regina v. Harrington* (1852), the deceased's throat and the red cotton straps of a nightcap had been cut with a sharp object. Matching red cotton fibers were found embedded in blood on the defendant's razor.[8]

In a New Orleans case (1858), a child was a victim of rape. Stains showed a mixture of blood and spermatozoa and included both blue and red wool fibers. The accused had been wearing a red wool flannel shirt over a bluish-gray woolen shirt.[8]

In *Regina v. Cass* (1860), a bloody knife was left at the crime scene. Wool fibers found mixed with the blood had a peculiar purple-brown color that corresponded exactly to those from the coat of the accused.[8,9]

In Littlejohn's case of infanticide (1875), a newborn was found with a thread tying the umbilical cord. Similar thread in the accused mother's possession helped establish the case.[9]

The most detailed and fascinating case encountered was the Vail case (1890). Dillon[1] discusses this case, but the original literature goes into much more detail, giving the full text of Frank L. James's presidential address to the American Society of Microscopists.[24] The case centers on a firearms shooting reconstruction that includes a microscopical soil comparison and examination of the scorching effects of the muzzle flash on wool, cotton, linen, and silk. James concluded that the shooting was accidental and the defendant was acquitted following a celebrated trial.

The last case on our list is *Regina v. Kerr* (1898). In this case an assault had occurred in a rural area, and the victim had worn a red wool petticoat. Mud stains were on both knees of the accused's trousers. The soil matched that from the crime scene area and red wool fibers matching the victim's petticoat were found in the mud stains.[8]

Soil and Miscellaneous Trace Evidence

The potential of other types of trace evidence also was recognized in the 1890s. Edgar and Johnston state:

> Particles of dirt should be preserved. They may have been gathered up in the spot where the crime was committed and thus furnish valuable corroborative testimony....
>
> The character of the earth on the man's boots or clothes may be found to correspond with that of the ground designated. Stains of paint, tar, oil, in one case fibres of red woolen underwear caught in the pubic hair, have served as additional means of identification.[25]

Tidy also expounds on the usefulness of stains and soil:

> In the case of dead bodies, all such details as smears of tar, paint, etc., either on the clothes or the body, should be recorded as likely to throw light on the identity of the person or help otherwise in assisting justice....
>
> It is most important to compare mud stains that may be found on the clothes of the accuser, with mud stains existing on the coat or trousers or other garments of the accused; and, further, to compare both with the earthy matter found at the precise spot where the assault was said to have been committed.[9]

Elsewhere, in considering bloodstain identification and the differentiation of other types of stains, Tidy emphasizes that stains are independently important as associative evidence and gives methods for the identification of grease, tar, pitch, tobacco, iron mold, paint, dyes, fruit stains, and flower stains.[9]

Table III lists cases involving miscellaneous trace evidence. The earliest of these is one of the more interesting. In *Regina v. Snipe and Others* (1852), the evidence showed that "some spots of mud on the boots and clothes of the prisoner, when examined microscopically, presented infusorial shells, and some rare aquatic vegetables, particles of soap and conferrae, and hairs from the seeds of groundsel."[8] Mud from a ditch close to where the body of the deceased was found showed the same microscopical appearances. Mud from all of the other ditches in the area was examined and found to be different, resulting in an opinion that the mud was from the ditch near the body.

In another early case of Taylor's (1853), soil on a woman's bonnet helped confirm her accidental death. Competing theories for her head injuries were a fall in the street or hammer blows.[8]

Three of the cases listed involved flour as trace evidence. Taylor cites an 1857 case where granules of wheat starch mixed with blood served as evidence in a murder trial.[8] In *Regina v. Steed* (1863), cited earlier, starch also was found on the accused's boots and corroborated testimony from witnesses who had seen the man earlier in a flour shop.[8,9] The third example is Rossin's 1867 rape case, where wheat and potato starch were found in a semen stain on a dress. The starch was found only in the stains and not on the dress itself. Similar stains were found on the accused's shirt. He used flour in his business and there was an open sack of flour at the foot of the bed where the alleged attack occurred.[8]

In the Moore case (1859), a woman's throat had been cut and small particles of steel were found in the spinal column. These were physically matched to the edge of the accused's razor.[8]

The Thidet case (1887) involved the drowning of a 15-year-old boy. Shoeprints led down the bank and into the water of a ditch. Sand grains from the suspect's trousers were matched to the ditch and to mud from the clothing of the suspect's daughter, who had served as an accomplice.[26]

Dillon cites an 1897 case where a microscopical glass analysis was conducted by one member of the jury who happened to be a glass worker.[1,27]

We conclude with some case work cited by Gross.[2] By 1906, when the English translation of Gross's 1904 work appeared, the role of the micro-

Table III. Case Examples of Miscellaneous Trace Evidence.

Year	*Case and Reference*
1852	*Regina v. Snipe and Others* (8, Vol. 1, pp 574, 575) Soil matching crime scene on suspect's boots.
1853	Taylor's accidental death case (8, Vol. 1, pp 532, 533) Soil on victim's clothing supporting case reconstruction.
1857	Taylor's wheat starch case (8, Vol. 1, p 575) Wheat starch in bloodstains on suspect's clothing.
1859	Moore case (8, Vol. 1, p 569) Physical match of metal fragment with suspect's razor.
1863	*Regina v. Steed* (8, Vol. 1, pp 562, 575; 9, Vol. 1, pp 238, 239) Flour on suspect's boots corroborating eyewitness testimony as to presence in a baker's shop.
1867	Roussin's rape case (8, Vol. 2, p 463) Wheat and potato starch matched to crime scene found in a semen stain on victim's clothing.
1887	Thidet case (*26*) Mud and sand from scene matched to clothing of two suspects.
1897	A glass case (*1*, pp 83, 84; *27*) Glass examined microscopically by a juror.
1898	*Regina v. Kerr* (8, Vol. 2, pp 467, 468) Fibers matching victim's clothing found on suspect's clothes, in mud stains matching crime scene.
~1900	Gross's cases (2, pp 212, 216) Environmental location of dusts and layered mud cases.

scopist was well recognized. In the use of forensics experts described by Gross, the microscopist is second only to the medical expert, and a full 29 pages describe the role of the microscopist (compared, for example, to 3 pages for the chemical analyst). Gross's work is beyond our cutoff date, but we may reasonably infer that the work he cited occurred during our period of interest. Gross emphasized the ability to characterize the surrounding environment from a dust sample and gave examples within his own experience of attributing the dust to a desert, a ballroom, a smithshop, a study, a locksmith, a miller, a school boy, and a chemist. He stated:

> All these examples are drawn from the author's own practice and in all of them, neither a determinate body nor a particular particle of a determinate body was being searched for; but the dust was collected for microscopic examination and in each case, new clues were found therefrom enabling the inquiry to proceed.[2]

Gross also cites two cases where the significance of trace evidence on boots was greatly enhanced by the careful examination of adherent mud:

> In both cases the mud on the boots of the accused was examined and in both cases two layers of mud were found separated from one another, in the first case by flour and in the second case by fine sand. In the former the accused had walked with muddy boots in the flour laying about the mill: in the latter he had also walked, first in the mud, then in the sand of the river bank, and then again in the mud.[2]

Summary

Our survey of forensic microscopy in the 1890s has demonstrated that there was an awareness of the evidential value of trace evidence materials. Many of the same types of materials that we examine today had received scientific scrutiny and had been introduced as evidence in court. One hundred years later, we find a greater variety of particles that can serve as trace evidence (synthetic fibers, polymers, and organic pigments, to name a few), but the use of trace evidence must necessarily begin with its recognition and careful microscopical examination. Indeed, it frequently appears that we are distracted by the presence of other types of evidence, such as fingerprints and bloodstains, where the value is macroscopical as well as highly discriminating. For some, the virtues of microscopical trace evidence are slipping from memory and what is easily achieved through careful casework seems mere fantasy. But trace evidence is there to be found and used as those who look know, and its presence and value were recognized in the 1890s.

Development of the Comparison Microscope

Although the development of the comparison microscope extends beyond 1900, the importance of this instrument warrants a brief discussion of its history and early development. (See Table IV for a summary of important dates.)

Émile M. Chamot summarized the development of the comparison microscope through 1915 in his textbook *Elementary Chemical Microscopy*.[28] Unfortunately, this historical review is omitted in his later work *Handbook of Chemical Microscopy*.[29] Although the *Handbook* continues to receive considerable exposure in the forensic sciences, we have remained ignorant on this point. Contemporary historical discussions have been given by Mathews,[30] Thornton,[31] and Leitz.[32]

The comparison bridge, allowing images from two microscopes to be viewed simultaneously, was developed in 1885 by Alexander von Inostranzeff, a Russian mineralogist.[28,31–34] Inostranzeff's application was the critical comparison of minerals by reflected light. He initially had used a camera lucida with partial field stops to combine the images of two microscopes. His new design employed two sets of totally reflecting prisms mounted so that one-half of the field of two separate microscopes would be reflected into a single eyepiece. In 1887 Van Heurck modified the design by allowing for a finer division between the two optical fields.[28,35]

This early work on comparison microscopy had escaped the attention of contemporary forensic scientists until Thornton[31] came across the original German article while conducting unrelated research. These optical devices allow images from two separate microscopes to be compared. We thus use the term "comparison eyepiece" or "comparison bridge" to describe them. Complete comparison microscopes incorporate the bridge optics directly onto a single microscope. The distinction can be readily appreciated by comparing Figures 1 and 2.

Document Comparison

The first published applications of comparison microscopy in forensic science were by Albert S. Osborn. Osborn developed the "color microscope",

Table IV. Development of the Comparison Microscope.

Year	*Development*
1885	Alexander von Inostranzeff, a Russian mineralogist, develops the comparison bridge after using a camera lucida to superimpose images from two microscopes.
1887	H. Van Heurck modifies Inostranzeff's optical design to allow for a finer division between the two optical fields.
1908	Albert S. Osborn develops the Color Microscope for comparison of inks and notes that there are many other applications to forensic document examination.
1910	Marshall D. Ewell independently develops a more elaborate color comparison microscope including transmitted light illumination and iris diaphragms in the objectives.
1911	W & H Seibert Optical Institute produces a commercial comparison microscope, mechanically similar to Ewell's, with a movable prism to adjust the proportion of the field of view occupied by each image.
1912	The Van Heurck comparison eyepiece is offered commercially by Bausch & Lomb Optical Company.
1913	Leitz develops a binocular comparison microscope consisting of two separate optical systems where each eye receives the image from one of the microscopes.
1914	A comparison eyepiece is offered commercially by Leitz.
1915	Émile M. Chamot reviews several comparison microscopes and eyepieces and emphasizes the utility of comparison microscopy for forensic science investigations.
1922	Émile M. Chamot applies comparison microscopy to the study of small arms primers. A monocular comparison microscope is offered by Leitz.
1925	Philip O. Gravelle constructs the first comparison microscope designed for forensic firearms examination and Calvin H. Goddard begins applying the technique to bullet and cartridge case comparisons. J. Howard Mathews special-orders a bullet comparison microscope from Bausch & Lomb.
1927–1930	Forensic laboratories develop their own comparison microscopes using commercially available comparison bridges. There is considerable innovation regarding the design of bullet and cartridge case holders. Commercial comparison microscopes designed specifically for firearms examinations become available.

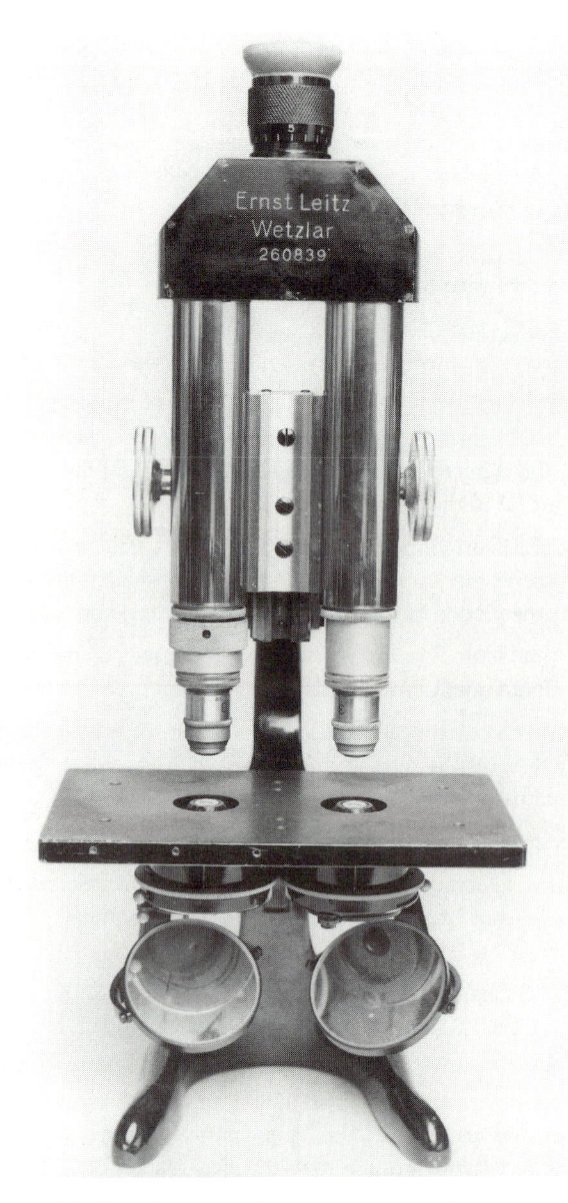

Figure 1. The Paul L. Kirk hair and fiber comparison microscope, manufactured circa 1922 by E. Leitz of Germany. Kirk used this microscope while teaching at the University of California at Berkeley. His interest in hair identification is well known, and this microscope would be well suited for his work. (Photograph courtesy of Tom Culbertson, Ventura County Sheriff's Department Regional Photographic Laboratory.)

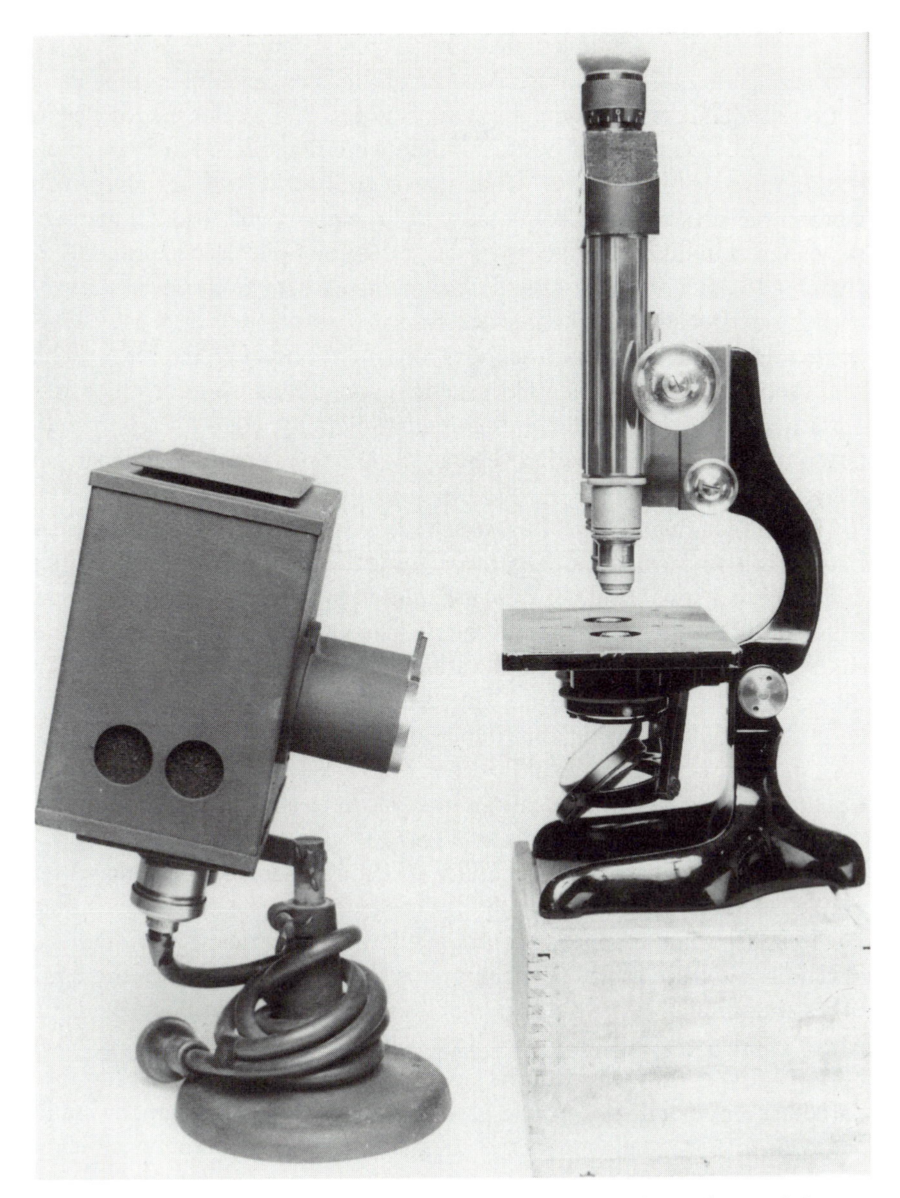

Figure 2. The Leitz comparison microscope with a Leitz lamp designed for use with this microscope. The lamp was not *used by Professor Kirk, but is shown here to illustrate how the microscope was used with this special twin beam light source. (Photograph courtesy of Tom Culbertson, Ventura County Sheriff's Department Regional Photographic Laboratory.)*

a comparison microscope designed for the critical comparison of ink colors. This was particularly important because of the change in color upon aging of iron-nutgall inks. Osborn published his notes on the color microscope in 1908 in the *Chicago Legal News*.[36] These were republished in 1910 in his book *Questioned Documents*.[37] The question of earlier publication of this method has arisen,[31] possibly because of his earlier publication of many of the chapters in his book. Osborn discusses his color microscope primarily in Chapter 18, "Ink and Questioned Documents", where he specifically attributes the portions of the chapter dealing with this topic to his earlier 1908 publication. The color microscope is also mentioned in Chapter 5, "The Microscope and Questioned Documents", an earlier version of which was published in 1903.[38] Although the comparing of ink color for age evaluation is mentioned in this earlier version, there is no reference to comparison microscopy.

The color microscope was designed by Osborn and constructed by Bausch & Lomb Optical Company.[30] It was used strictly for incident light work and had "two parallel tubes surmounted by enclosed reflecting prisms which bring the rays of light from the two objects into juxtaposition so that each image occupies one-half of the field as seen under one eyepiece".[37] Standardized colored Lovibond Tintometer glasses could be inserted into either body tube, allowing one to measure "the most delicate distinctions, both in tint and depth of color". Although it was designed for comparisons, Osborn noted that "many occasions arise in the examination of questioned documents when other uses can be made of such an instrument".[37] Mitchell noted in particular the utility for documenting color change reactions used in ink analysis.[39]

In 1910, Marshall D. Ewell independently constructed a comparison microscope for use as a color comparator.[40] The construction of the body tubes was similar, but Ewell's design was more elaborate, incorporating paired condensers for transmitted light, a fine focus, and iris diaphragms in the objectives. As with Osborn's microscope, the body tubes accommodated the Lovibond Tintometer glasses. A Spencer Lens Company stand was used, and the remainder of the comparison microscope was constructed in Ewell's shop. The editor of the *Journal of the Royal Microscopical Society* compared Ewell's instrument to the earlier works of Inostranzeff and Van Heurck:

> The general idea recalls Inostranzeff's comparison chamber or microscopic comparer.... This idea was somewhat improved upon by van Heurck. It will be noticed that Dr. Ewell's design, which has

> been independently evolved, is an advance in several respects. It requires only one Microscope; it secures par-focality of objectives, and, owing to the use of only two prisms [additional] loss of light... is avoided.[40]

In 1911, a comparison microscope mechanically similar to Ewell's was produced by W & H Seibert's Optical Institute in Wetzlar on the suggestion of Wilhelm Thörner of Osnabrück.[28,32] This instrument took Inostranzeff's original optical design but incorporated a movable prism design so that one could adjust the relative proportion of the field of view to be occupied by each image.

The Van Heurck comparison eyepiece was offered commercially by Bausch & Lomb Optical Company in 1912,[28] and a comparison eyepiece based on the Inostranzeff optical system was made by Leitz in 1914.[28,32] Zeiss had developed a comparison eyepiece by 1925.

A *binocular* comparison microscope, mechanically similar to the Ewell design, was offered by Leitz in 1913.[32] Each body tube had its own reflecting, erecting prism that rotated to adjust interpupillary distance. Each eye received one image, and sliding stops in each eyepiece allowed portions of the field to be blocked.[28] Images could thus be juxtaposed, superimposed, or combined in any proportion. Chamot achieved excellent results with this microscope and, in 1915, proposed its use for forensic applications:

> The fields are flat, brilliant, and with careful illumination and adjustment and a little practice most excellent results can be obtained. The instrument is adapted to all problems involving an exact comparison of size, structure or symmetry of microscopic objects, especially where the structure is so intricate as to render comparison and interpretation with the single compound microscope exceptionally difficult without recourse to photography. The value of the instrument in all problems of forensic chemical microscopy is evident.[28]

Chamot also evaluated the W & H Seibert comparison microscope:

> The instrument may be employed with polarized light, thus affording exceptional opportunities for exact comparisons in the search for food adulterants and in microchemical analysis. Since in this instrument we have a single ocular yielding a divided field, it is possible to obtain photomicrographs, half the area of the circle in the negative obtained being the image of one preparation, the other half that of the second preparation.[28]

In 1922, Leitz introduced a monocular comparison microscope, using the Seibert design,[32] which is reviewed in Chamot and Mason's *Handbook of Chemical Microscopy*.[29] This microscope is shown in Figures 1 and 2. Another binocular comparison microscope, looking somewhat like a stereomicroscope with wings, was later produced by Leitz and marketed as a "photographic binocular microscope". This microscope produced superimposed images.[41]

Firearms Examination

The first application of comparison microscopes to firearms examinations was by Émile M. Chamot in his 1922 monograph *The Microscopy of Small Arms Primers*.[42] The microscope used by Chamot was made by Bausch & Lomb specifically for this study and apparently included special mechanical stages not illustrated in the monograph.[30] The first to apply comparison microscopy to forensic firearms identification (matching projectiles and cartridge casings to firearms) were Calvin H. Goddard and Philip O. Gravelle at the Bureau of Forensic Ballistics in New York. Gravelle designed the instrument in 1925 and left the bureau shortly thereafter.[43–46] He had earlier used a comparison microscope for comparing weave patterns in cloth.[30] Goddard applied the instrument to casework and popularized the methods[30] and, following his work on the St. Valentine's Day Massacre, he left to become the director of the new Scientific Crime Detection Laboratory in Chicago.[45]

Gravelle's instrument used a "Zeiss comparison bridge, Spencer tubes, Leitz eyepieces, Bausch & Lomb objectives, and bullet mounts constructed to order by the Remington Arms Company".[31] A similar instrument was used by Edward C. Crossman, who began as Goddard's West Coast representative (see Figures 3 and 4).

Being aware of Chamot's work, J. Howard Matthews special-ordered a bullet comparison microscope from Bausch & Lomb in 1925.[30] Over the next five years a variety of prototype instruments were constructed by firearms examiners, with different bullet and cartridge case holders and different combinations of commercially available and improvised apparatus.[47–50]

The claim that Gravelle's work was the earliest application to firearms identification has been occasionally disputed, most notably in Sir Sidney Smith's popular autobiography.[51] Smith makes reference to an improvised comparison microscope that was employed prior to using "Waite's" design, specifically citing the period of 1919 to 1922, as well as his subsequent

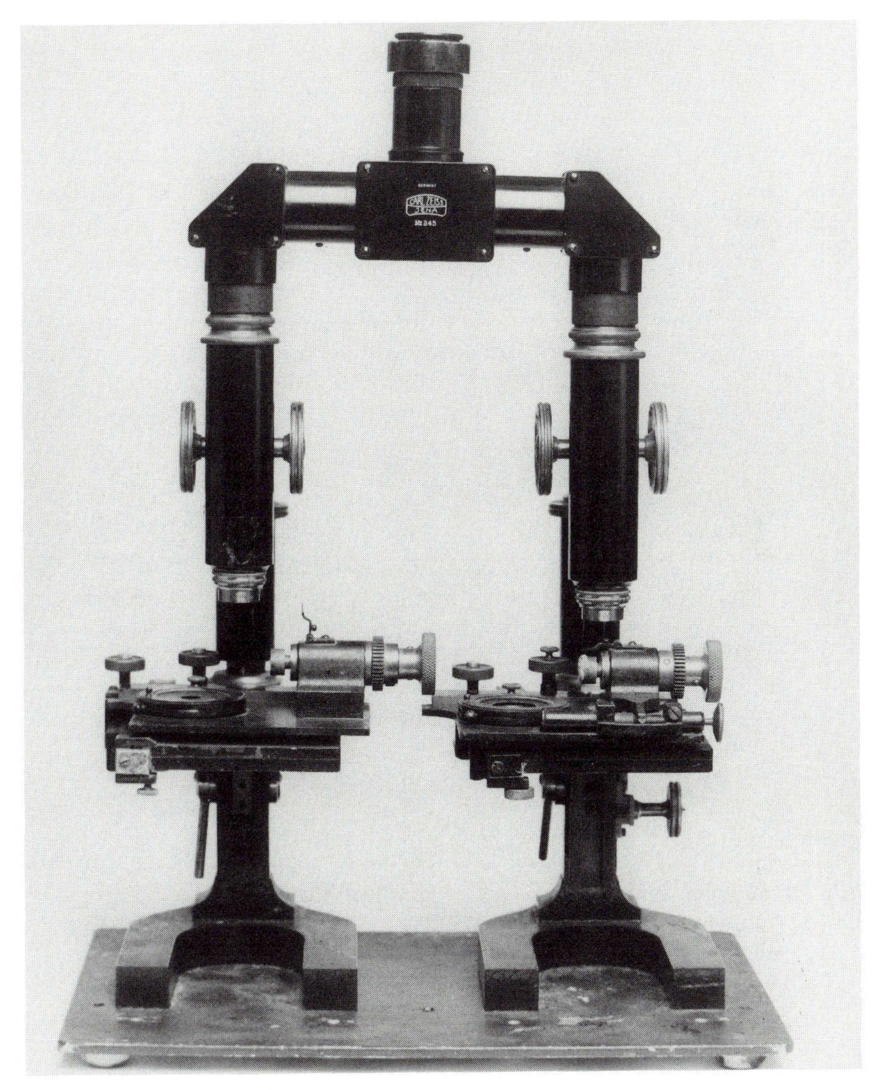

Figure 3. The Captain Edward C. Crossman bullet comparison microscope, circa 1926, which he used in his work in Los Angeles until his death in 1939. Crossman is shown with this microscope in the frontispiece of his book The Book of the Springfield *(ref. 65). Crossman worked for the Los Angeles County sheriff and district attorney as well as other law enforcement agencies during the 1920s and 1930s as an expert witness in firearms identification. He began his work in this area as the West Coast representative of Calvin Goddard. (Photograph courtesy of Tom Culbertson, Ventura County Sheriff's Department Regional Photographic Laboratory.)*

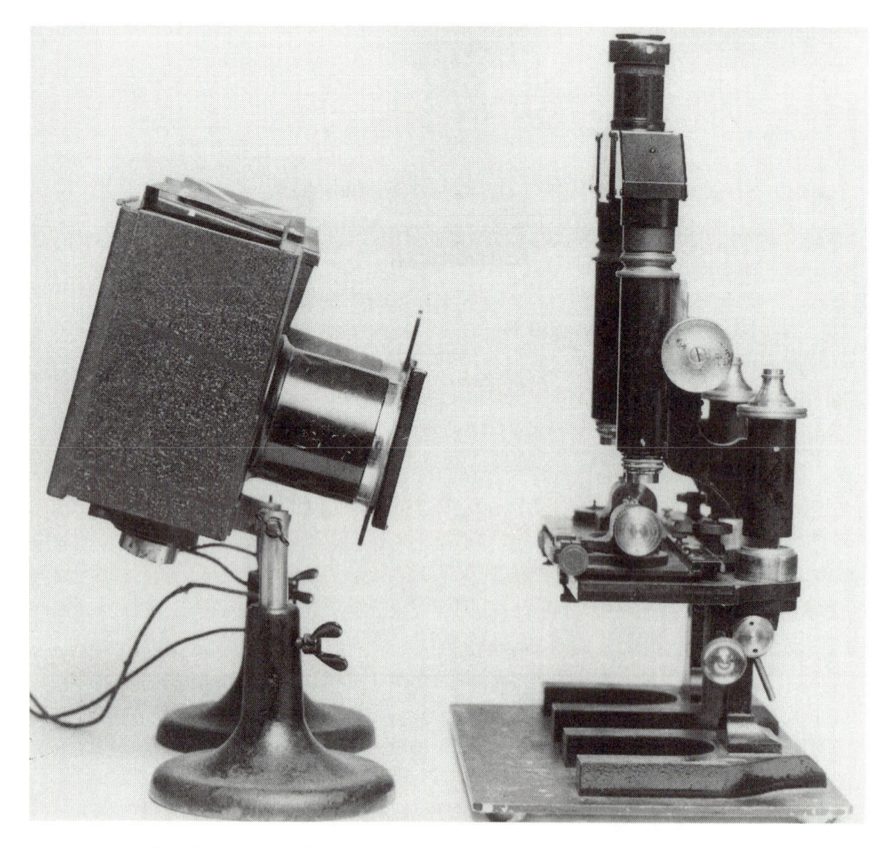

Figure 4. The Captain Edward C. Crossman comparison microscope with its Bausch & Lomb light source, which he used with the microscope throughout his career. Note the bullet holders illustrated here and in Figure 3. (Photograph courtesy of Tom Culbertson, Ventura County Sheriff's Department Regional Photographic Laboratory.)

analysis of evidence in the murder of Sir Lee Stack Pasha in 1925. This claim appears to be unfounded. In his contemporaneous account of these investigations, Smith[52] makes no mention of a comparison microscope. In the 1931 book, *Recent Advances in Forensic Medicine* (coauthored by Sidney Smith and John Glaister, Jr.) a comparison microscope is pictured with the note, "This instrument was first described by Goddard and Waite".[53] Furthermore, Lucas, who attributes precedence to Goddard (corrected in a later edition to Gravelle) was actually in Egypt, at the time, working on these same cases.[54,55]

In the fall of 1929, after the establishment of the Scientific Crime Detection Laboratory at Northwestern University, Calvin Goddard made an 80-day trip to Europe to study prevailing practices in scientific crime detection. Following his trip, he reported that four laboratories he visited (London, Cairo, Lyon, and Edinburgh) were using comparison microscopes in their firearms work, "patterned after the one devised by Gravelle of South Orange, N.J., in 1925 and used by me constantly since".[46] Goddard describes the microscopes he encountered:

> Of the two adaptations that I had an opportunity to examine personally, the first (at London) was badly designed (although its author had studied my own instrument in my laboratory in New York), as it places the bullet images with their axes in a vertical, instead of a horizontal plane, and has rigid lighting fixtures, permitting no modifications according to conditions, while the second (at Lyon), was hopelessly wrong in design, so wrong as to render a bullet identification with it absolutely impossible (since it fuses the images nose to nose, or base to base, instead of employing the nose of one bullet and the base of the other). The microscope had been prepared for the commercial market by a student working in Locard's laboratory, from a study of an illustration in a reprint sent him by me in 1927. Of the third (at Cairo), I had an opportunity to study an illustration only. It too, was patterned after a photograph reproduced in one of my articles and appears to be correct in principle and detail. The fourth (at Edinburgh), presumably copies that in Cairo, having been installed by the same man, Dr. Sidney Smith.[46]

With regard to the London microscope, Goddard is referring to Robert Churchill, who traveled to the United States in 1927, met with prevailing experts (including Goddard), and returned to help design Watson's Comparison Microscope Mark II.[56] As for the microscope in Lyon, the "student" in Locard's laboratory was Harry Söderman. In his 1928 work, *L'Expertise des Armes à Feu Courtes*,[49] Söderman makes reference to the "*méthode de Goddard et Waite*", citing the 1925 *Army Ordnance* article,[43] and illustrates the comparison microscope he had constructed by J. Gambs.[49] This microscope was later marketed as the "Hastoscope",[48,57] to which Goddard refers. Kraft gives a much more favorable review of the Hastoscope, stating "before this instrument was set up, we used the comparison eyepiece and two microscopes as described by Goddard.... For purposes of bullet examination the hastoscope was undoubtedly a great step forward."[41] It is interesting that in the French literature, in spite of Söderman's explicit credit to Goddard,[49] Söderman is given full credit for the development of the comparison microscope: "En 1925 a été fabriqué, sous les directives de Söderman, le premier *microscope de comparaison*, appelé hatoscope [sic], ancêtre des appareiles modernes".[58]

It would appear, however, that Goddard's comments regarding the other comparison microscopes available in 1930 were not justified. Kraft postulates that during Goddard's trip he either did not visit the places in question or that he wasn't shown everything to be seen, stressing that Metzger's comparison microscope[47] was indeed very usable.[48] Kraft also suggested that it was the early photographs of the Hastoscope that caused Goddard to infer that it was "hopelessly wrong in design".[48]

In any case, the forensic firearms comparison microscope developed rapidly, and by 1931 there were at least five commercially available instruments.[32, 56–57,59–60]

Trace Evidence Examination

Thus far, this review covers the application of comparison microscopes to document and firearms examination. It is less clear when comparison microscopes were generally accepted in forensic laboratories as part of trace evidence examination. Apart from Chamot's observation in 1915,[28] there is little mention of comparison microscopes in general use in forensic laboratories. Locard, in 1923, discusses Osborn's methods using a comparison microscope for critical comparison of inks, but there is no reference to comparison eyepieces or applications to either trace evidence or firearms examination.[61] Söderman and O'Connell, in 1935, state that "the instrument was first used to identify bullets by the Americans Gravelle and Waite",[62] but again there is no mention of trace evidence applications.

The first use of a comparison microscope by the FBI was apparently in 1934 when Charles Saylor (National Bureau of Standards) and Harry Pickering (FBI) adapted an inverted demonstration eyepiece so that it could simultaneously view the work product of two typewriters.[63] This was the year when the technical functions of the FBI were just beginning to expand.[1] A comparison microscope must have been obtained fairly quickly after that, as photographs in Söderman and O'Connell's *Modern Criminal Investigation* (1935) show an instrument in use at the FBI Laboratory.[62] Three years later Pickering, then a research chemist in Florida, described how comparison microscopes were used to "compare two of anything such as bullets, shells, splinters, cloth, hairs, etc."[64] It is reasonable to assume that once the comparison microscope became available in forensic laboratories, it was used for other applications when needed and no need was felt for specific documentation of this effort.

Outside of forensic science, the comparison microscope was not readily assimilated into analytical laboratories. As late as 1958, Needham noted, "unfortunately, only the criminologist has taken full advantage of the comparison microscope.... In research and industrial laboratories... the comparison microscope is practically unknown."[44]

Acknowledgments

Special thanks are given to Duayne Dillon for his help in locating reference material, to Tom Culbertson of the Ventura County Sheriff's Department for the photographs in Figures 1–4, and also to the editor for his exceptional patience. The microscopes and lamps depicted in Figures 1–4 are part of the Paul M. Dougherty collection.

References

1. Dillon, D. J. D.Crim. Dissertation, University of California at Berkeley, 1980; pp 2, 51, 52, 83, 84,174–177.
2. Gross, H. *Criminal Investigation;* Translated by Adam, J. and Adam, J. C.; Krishnamachari: Madras, India, 1906; pp 212, 216.
3. Griffith, J. W.; Henfrey, A. *The Micrographic Dictionary*, 4th ed.; Van Voorst: London, 1883.
4. Hassall, A. H. *Food: Its Adulterations, and the Methods for Their Detection;* Longmans, Green: London, 1876; pp 857–858.
5. *Microscopic-Analytical Methods in Food and Drug Control;* Harris, K. L.; Reynolds, H. L., Eds.; Food and Drug Technical Bulletin No. 1; Food and Drug Administration: U.S. Government Printing Office: Washington, DC, 1960; pp 2–4.
6. Wormley, T. G. *Micro-Chemistry of Poisons*, 2nd ed.; Lippincott: Philadelphia, PA, 1885.
7. Blyth, A. W. *Poisons: Their Effects and Detection;* Wood: New York, 1885.
8. Taylor, A. S. *The Principles and Practice of Medical Jurisprudence*, 4th ed.; Stevenson, T., Ed.; Churchill: London, 1894; Vol. 1, pp 528–533, 542, 561–567, 574–575; Vol. 2, pp 461–472, 594–601. Alternatively, see the American edition: Taylor, A. S. A *Manual of Medical Jurisprudence*, 12th ed.; Stevenson, T.; Bell, C., Eds.; Lea: Philadelphia, PA, 1897.
9. Tidy, C. M. *Legal Medicine;* Wood: New York, 1882; Vol. 1, pp vi, 129, 169–175, 185, 186, 200–206, 222–244; Vol. 3, p 135.
10. *Medical Jurisprudence, Forensic Medicine and Toxicology;* Witthaus, R. A.; Becker, T. C., Eds.; Wood: New York, 1894–1896.
11. Ballowitz, E. *Inter. Monatsschrift f. Anat. u. Phys.* **1890**, 7. (Ballowitz published extensively on the topic of spermatozoa morphology.)

12. Formad, H. F. *Mammalian Blood, with Special Reference to the Microscopical Diagnosis of Blood Stains in Criminal Cases*; Hummel: Philadelphia, PA, 1888.
13. Moeller, J. *Pharmakognosticher Atlas*; Springer: Berlin, 1892.
14. van Ledden Hulsebosch, M. L. Q. *Makro- und Mikroskopische Diagnostic der Menschlichen Exkremente*; Springer: Berlin, Germany, 1889.
15. Wood, E. S. In *Medical Jurisprudence, Forensic Medicine and Toxicology*; Witthaus, R. A.; Becker, T. C., Eds.; Wood: New York, 1894; Vol. 2, pp 3–82.
16. Park, R. In *Medical Jurisprudence, Forensic Medicine and Toxicology*; Witthaus, R. A.; Becker, T. C., Eds.; Wood: New York, 1894; Vol. 1, pp 591–626.
17. Pfaff, E. *Das Menschliche Haar, etc.*; Leipzig, 1866 (from 8, Vol. 1, p 173).
18. Rosse, I. C. In *Medical Jurisprudence, Forensic Medicine and Toxicology*; Witthaus, R. A.; Becker, T. C., Eds.; Wood: New York, 1894; Vol. 2, pp 491–518.
19. Wood, E. S. In *Medical Jurisprudence, Forensic Medicine and Toxicology*; Witthaus, R. A.; Becker, T. C., Eds.; Wood: New York, 1894; Vol. 2, pp 83–96.
20. Rosse, I. C. In *Medical Jurisprudence, Forensic Medicine and Toxicology*; Witthaus, R. A.; Becker, T. C., Eds.; Wood: New York, 1894; Vol. 1, pp 383–435.
21. *American Monthly Microscopical Journal* **1889**, *11*, 187–188.
22. Cook, J. G. *Handbook of Textile Fibers*, 4th ed.; Merrow: Watford, England, 1968; Vol. 2, pp 3–6.
23. Lamb, D. S. In *Medical Jurisprudence, Forensic Medicine and Toxicology*, Witthaus, R. A.; Becker, T. C., Eds.; Wood: New York, 1894; Vol. 1, pp 705–792.
24. James, F. L. *American Monthly Microscopical J.* **1891**, *13*, 221–229.
25. Edgar, J. C.; Johnston, J. C. In *Medical Jurisprudence, Forensic Medicine and Toxicology*; Witthaus, R. A.; Becker, T. C., Eds.; Wood: New York, 1894; Vol. 2, pp 413–490.
26. Locard, E. *American Journal of Police Science* **1930**, *1*, 496–514.
27. *American Monthly Microscopical Journal* **1897**, *18*, 31 (from *1*, pp 83, 84).
28. Chamot, E. M. *Elementary Chemical Microscopy*; Wiley: New York, 1915; pp 23–28.
29. Chamot, E. M.; Mason, C. W. *Handbook of Chemical Microscopy*; Wiley: New York, 1931; Vol. 2, pp 68–70.
30. Mathews, J. H. *Firearms Identification*; Thomas: Springfield, IL, 1962; Vol. 1, pp 36–43.
31. Thornton, J. I. *Association of Firearms and Toolmarks Examiners Journal* **1978**, *10*.
32. *Leitz Bulletin for the Forensic Laboratory*; Leitz: New York, 1986; Nr. 5/2.86.
33. Inostranzeff, P. A. V. *Neues Jahrbuch fur Mineralogie, Geologie, und Palaeontologie* **1885**, *2*, 94–96.
34. Anonymous. *Journal of the Royal Microscopical Society* **1886**, *6(series II)*, 507–509.
35. Anonymous. *Journal of the Royal Microscopical Society* **1887**, 463–464.
36. Osborn, A. S. *Chicago Legal News* **1908**, *11(23)*, January. (Citation taken from reference 37).
37. Osborn, A. S. *Questioned Documents*; Lawyers Co-Operative: Rochester, NY, 1910; pp 70, 74–75, 102–104, 355–362.
38. Osborn, A. S. *Journal of Applied Microscopy and Laboratory Methods* **1903**, 6, 2637–2643.
39. Mitchell, C. A.; Hepworth, T. C. *Inks: Their Composition and Manufacture*, 3rd ed.; Griffin: London, 1924; pp 177–181.
40. Ewell, M. D. *Journal of the Royal Microscopical Society* **1910**, *Part 1*, 14–16.
41. Kraft, B. *American Journal of Police Science* **1931**, *2*, 409–418. (Originally appearing in *Archiv für Kriminologie* **1931**, 88, 211.)

42. Chamot, É. M. *The Microscopy of Small Arms Primers*. Cornell: Ithaca, NY, 1922.
43. Goddard, C. H. *Army Ordnance* **1925,** 6, 197.
44. Needham, G. H. *The Practical Use of the Microscope*; Thomas: Springfield, IL, 1958; pp 83–87.
45. Dougherty, P. M. *Association of Firearms and Toolmarks Examiners Journal* **1991,** *23*, 900–902.
46. Goddard, C. *American Journal of Police Science* **1930,** *1*, 13–37.
47. Mezger, O. *Deutsche Zeitschrift für die gesamte gerichtliche Medizin* **1929,** *13*, 377.
48. Kraft, B. *American Journal of Police Science* **1931,** *2*, 52–66 and 125–142. (Originally appearing in *Archiv für Kriminologie* **1930,** *87*, 133.)
49. Söderman, H. *L'Expertise des Armes à Feu Courtes*; Desvigne: Lyon, France, 1928; pp 68–70.
50. Derome, W. *American Journal of Police Science* **1930,** *1*, 216–223.
51. Smith, S. *Mostly Murder;* Harrap: London, 1959; pp 98–114.
52. Smith, S. *The Police Journal* **1928,** *1*, 411–422. Originally published in *British Medical Journal* **1926,** *i*, 8–10.
53. Smith, S.; Glaister, J., Jr. *Recent Advances in Forensic Medicine;* Churchill: London, 1931; pp 51–53.
54. Lucas, A. *Forensic Chemistry and Scientific Criminal Investigation*, 2nd ed.; Longmans, Green: New York, 1931; pp 217–218.
55. Lucas, A. *Forensic Chemistry and Scientific Criminal Investigation*, 3rd ed.; Longmans, Green: New York, 1935; p 257.
56. Hastings, M. *The Other Mr Churchill;* Dodd, Mead: New York; pp 126–140.
57. Gambs, J. (advertisement). *Revue Internationale de Criminalistique* **1931,** *3(8)*, inside front cover.
58. Gayet, J. *Manuel de Police Scientifique;* Payot: Paris, 1961; p 106.
59. Bausch and Lomb Optical Company (advertisement). *American Journal of Police Science* **1931,** *2*, 559.
60. Spencer Lens Company (advertisement). *American Journal of Police Science* **1931,** *2*, 191.
61. Locard, E. *Manuel de Technique Policière;* Payot: Paris, 1923; pp 179–183.
62. Söderman, H.; O'Connell, J. J. *Modern Criminal Investigation;* Funk and Wagnalls: New York, 1935; p 188, 208, 409.
63. Saylor, C. P. *SPIE* **1977,** *104*, 31–37.
64. Pickering, H. S. *The Laboratory Detective;* Pickering: Miami, FL, 1937; p 91.
65. Crossman, E. C. *The Book of the Springfield;* Small-Arms Technical Publishing Co.: Marines, NC, 1932.

7

Milestones of Forensic Toxicology

Marina Stajic

"The forcible administration of poison is by no means a new thing in criminal annals," remarked Sherlock Holmes to Dr. Watson while investigating a murder case at the very beginning of their partnership in 1881.[1] An understatement, indeed. Intentional and accidental poisonings were certainly not a novelty at the time. Moreover, postmortem toxicology had already begun to play an important role in the criminal investigation process.

Early History of Poisons and Poisoning

Many writings from ancient civilizations, including Mesopotamia, Egypt, India, and Sumeria, indicate familiarity with toxic substances of vegetable, mineral, and animal origin. The oldest written document on the subject seems to be an Egyptian scroll dating back to about 1500 B.C. and purchased in 1862 by German Egyptologist George Ebers, who published the facsimile of it as *The Ebers Papyrus, the hermetic book of Egyptian medicaments, in hieratic script.*[2] Fifteen years later the papyrus was translated by Joachim.[3] The effects of many naturally occurring toxins were known in ancient Greece. Democritus, who by today's standards would be considered a biochemist, lived in the ancient town of Abdera and was thought to be crazy by his fellow citizens for spending his time studying plants (including their toxic effects) and dissecting animals. Hippocrates, who himself was interested in plant and food poisonings, was actually sent to examine Democritus and found him to be one of the wisest men of the era. This was probably the beginning of the working relationship between physicians and applied chemists.

The familiarity with toxic substances in the Greek civilization is evident by the use of poisons as a means of execution, the most famous case being that of Socrates, who was forced to drink a hemlock solution. The ancient Romans continued this practice, just one of many they took over from the Greek culture. The earliest recorded criminal poisoning was in the Roman Empire in 331 B.C. "Live by the poison, die by the poison" was the expression applied to a group of women who were ordered to drink the same preparation they had used on their victims. The first law against poisoning was enacted in the Roman Empire in 82 B.C.[4] The law notwithstanding, poisons were a popular weapon among ancient Romans and provided plenty of job opportunities in the field of food testing, as well as the development of antidotes. Arsenic, readily available as a component of lead or tin ore, became a most popular poison in Roman days and persisted as the "king of poisons" well into the nineteenth century, having caused the ultimate demise of many famous and infamous persons.

The art of poisoning, like other forms of art, blossomed during the Renaissance and post-Renaissance. Motives for disposing of unwanted people did not differ much from today's—greed, jealousy, fear, lust for power—but the art of administering poison was subtle. Rings with secret compartments, swords and knives with hidden recesses, poisoned letters and handkerchiefs, even lips delivering a true kiss of death added a certain touch of romance to the grim enterprise. Poisoning societies were formed and even family businesses started. The Medici and Borgia families became better known for their poisonous endeavors rather than any of their many historic ones. France became another arena for poison application and romanticized arsenic as *poudre de succession* (inheritance powder).

As popular as arsenic was, it was by no means the only toxic preparation used. The reign of Louis XIV saw an enormous number of deaths by poison. Catherine Monvoisin, known as La Voisin and the most famous poisoner of those times, was just one of many to practice the very lucrative profession of using poison as "the sole solution to most family problems". Although a number of men practiced the profession of poisoning, it appears that the majority of "professionals" in the field were women. Not everyone got away with murder by poison (La Voisin ultimately was burned at the stake), but convictions were based primarily on circumstantial evidence and proof of malicious intent. In 1781, J. J. Plenck made a revolutionary statement in his book *Elementa Medicinae et Chirurgiae Forensis*[5] that "the only certain sign of poisoning is the botanical character of a vegetable poison or the chemical identification of a mineral poison found in the body". Easier said than done. The scientific tools for such examinations were simply not available. It took over 30 years to even begin to apply this approach in practice and to begin using principles of toxicology to investigate rather than commit the crime.

Developments in Toxicology, 1800–1850

Detecting Arsenic

The theoretical aspects of toxicology began to emerge at the beginning of the nineteenth. The pioneer in changing the approach to the study of toxic substances was Mathieu Joseph Bonaventure Orfila. Orfila left his native Spain and moved to Paris at an early age, devoting his efforts to the study and compilation of existing data and verification of experiments performed by his colleagues. In 1814, at the age of 26, he published the first major text on toxicology under the impressive title *Traité des poisons tirés des règnes minéral, végétal et animal ou Toxicologie Générale.*[6] The book contained the first classification of poisons as corrosives, astringents, acrids, stupefying or narcotics, narcotico-acrids, or putrefaciants. The book quickly gained international importance, and Orfila became the most renowned medicolegal expert of the time. A good deal of Orfila's work dealt with arsenic. For example, by administering arsenic to dogs, he showed that the substance was distributed to all parts of the body, which disputed Fonatana's theory from the end of the eighteenth century that poisons act on one type of body tissue alone.

The method for detection of small amounts of arsenic was developed by English chemist James Marsh[7] in 1836 and became the first analytical methodology used in toxicology to be presented in a criminal trial.

> Fortunately, the investigation of poison murders has recently been revolutionized by advances in the science of chemistry. Probably the defendant would not stand before this court at this moment had not science given us the means to prove the presence of poi-

> son where hitherto it could not be detected in the very body of the victims.[8]

These words, spoken by Decous, the public prosecutor in Brive, France, introduced forensic toxicology to the courts of law in 1840. The defendant, Marie Lafarge, was a young woman being tried for murdering her husband. All the circumstances surrounding the death of Charles Lafarge pointed to arsenic as the lethal weapon and Marie Lafarge as the culprit. Yet there was a matter of scientifically proving the presence of poison in the body.

The defense attorney, Paillet, acted as his colleagues have continued to do since: he contacted a forensic expert, the dean of chemistry and toxicology of the times, Orfila. Orfila wrote a criticism citing circumstantial evidence and failure to use the Marsh test as the latest analytical methodology for arsenic. The prosecution in turn engaged its experts to analyze the stomach contents by Marsh's method; to everyone's surprise no arsenic was detected. Orfila's previous experiments, however, had showed that there are cases where arsenic that cannot be detected in the stomach is still detectable in other tissues. Organs from the exhumed body, including the liver, spleen, lungs, heart, intestines, and brain, were analyzed, again with negative results. Finally, the prosecution experts analyzed the beverages and food administered to the victim and found these to contain large amounts of arsenic.

At this point the prosecutor insisted on summoning Orfila as the ultimate authority on the subject. Orfila formed a team of all the participating experts, and for two days they proceeded to reanalyze the evidence. The result was the dramatic and historic statement by Orfila that established toxicology as a forensic science:

> I shall prove, first, that there is arsenic in the body of Lafarge; second, that this arsenic comes neither from the reagents with which we worked nor from the earth surrounding the coffin; also, that the arsenic we found is not the arsenic component which is naturally found in every human body.[8]

Thus the testimony of toxicologists, for the first time in history, became the final determining factor in the jury's verdict of guilty. In addition, Orfila brought up two important points during his testimony. First, he explained that the inexperience of the analysts with the new methodology was the cause of the initial false negative results. Second, when asked whether the detected amount of arsenic was sufficient to indicate murder by poisoning, he replied that the amount cannot be considered by itself but only as a part of the whole picture, including other facts of the case, such as the medical symptoms prior to death, the opportunity to obtain the poison, and the existence of poisoned food and beverages. These considerations remain important to this day. The most sophisticated analytical methodology is good only when properly applied, so categorical statements regarding the significance of toxicologic findings should be avoided. Toxicologic results serve as guidelines for interpretation. The opinions are based on examination of all available evidence in the case. Orfila rightfully deserves the title of "father of forensic toxicology".

Once scientists were able to detect arsenic in body organs, methods continued to be developed to detect other metals and minerals, such as antimony, lead, mercury, bismuth, phosphorus, sulfur, and others.

Detecting Vegetable Alkaloids

Vegetable alkaloids were another story. Their effects (in small doses as medicine, in large doses as poisons) had been known since antiquity, but because of their chemical nature they were not easy to isolate from biological tissues with available methodologies. F. W. Serturner isolated morphine from opium as early as 1805. This was followed by isolation of strychnine from nux vomica in 1819, quinine from the bark of the cinchona tree in 1820, coniine from hemlock in 1826, nicotine from tobacco in 1828, and atropine from belladonna in 1833. Arsenic and other metals lost their appeal, as they were no longer undetectable poisons, and the alkaloids found their way into the hands of poisoners. Prosecutors had to rely once again on circumstantial evidence. As a prosecutor remarked in Paris during the trial for an alleged murder by morphine in 1823:

> Since plant poisons leave no trace behind, murder by such poisons is not punishable.... Henceforth let us tell would-be poisoners: Do not use arsenic; do not use metallic poisons, for they leave traces. Use plant poisons.... Fear nothing; your crime will go unpunished. There is no corpus delicti because it cannot be found.[9]

He won the conviction in the case, but many poisonings with alkaloids continued to remain undetected. Tests were available for detecting these substances in their pure form, but they continued to elude detection during postmortem examinations. The first step in toxicologic analysis, the isolation of the chemical from the tissue matrix, was the stumbling block. Even Orfila's experiments were without success, and he began to believe the isolation of alkaloids from tissues to be an impossible task.

In 1850, another murder initiated a new period in forensic toxicology. The scene of the crime was the Château de Bitremont in Belgium, the victim Gustave Fougnies, whose sister and accomplice to the crime had married the perpetrator, Count Hypollite de Boccarmé. The count had been impatiently waiting for his ailing brother-in-law to die and leave his fortune to his sister. Fougnies decided to get married instead and thereby signed his death sentence. His visit to the castle of his sister and brother-in-law ended with him dying, supposedly of a stroke. The autopsy showed no sign of brain damage, but the mouth, throat, tongue, and stomach had distinct chemical burns. Specimens were submitted for analysis to the Brussels laboratory of Jean Servais Stas, a student of Orfila.

Stas recognized that the burns had not been produced by sulfuric acid, as initially thought, and he proceeded for almost three months to search for the fatal agent. His tireless efforts, combined with a couple of lucky coincidences, resulted in the isolation of nicotine from all the organs. It turned out that the count had extracted nicotine from tobacco and force-fed it to his victim. The toxicologic evidence presented by Stas led to the conviction and execution of Count de Boccarmé.[9,10]

The organs received by Stas had been preserved in alcohol acidified by the vinegar which the assailants poured down the victim's throat after his death. Nicotine, being soluble in both alcohol and water, was thus in the alcoholic solution. Stas was able by repeated filtration and evaporation to isolate the drug into a concentrated solution and, by making that solution basic, to extract it into a solvent immiscible with water. He happened to choose ether, an excellent solvent for extraction of alkaloids. The residue left upon evaporation of the ether contained nicotine. It was easy at that point to apply available tests to prove the presence of the isolated drug. The method Stas discovered, thanks to his hard work and to chance, enabled future toxicologists to isolate all known alkaloids. The Stas method,[11] modified by Otto[12] five years later, became the major isolation technique in forensic toxicology for analysis of nonvolatile organic substances. In modified versions, it is still in use today.

Developments in Toxicology, 1850–1950

Forensic toxicology began to develop in the United States in the second half of the nineteenth century. Among numerous contributions to the literature during this period were *Elements of Medical Jurisprudence* by Beck and Beck[13] in 1863; *Microchemistry of Poisons* by Wormley[14] in 1867; *A Manual of Toxicology* by Reese[15] in 1874, followed by *The Text Book of Medical Jurisprudence and Toxicology* by the same author[16] in 1884; *Forensic Medicine and Toxicology* by Hemming[17] in 1889; and *Medical Jurisprudence, Forensic Medicine and Toxicology* (four volumes) by Witthaus and Becker[18] in 1894 and 1896.

Toxicologic evidence played crucial roles in several murder trials during the late 1800s. The most prominent name in the field during that time was that of Rudolf Witthaus, professor of chemistry and physiology at New York University. Toxicologic analysis of human organs was still relatively rare, in part because of the inefficient case investigations under the coroner system. The search for poisons was often an afterthought, and toxicologists had to deal with the additional burden of analyzing exhumed organs.

Massachusetts was the first state to establish a medical examiner's system in 1877, requiring the examiners to be qualified pathologists. (It took another 45 years to establish a toxicology laboratory within the same system.) New York City replaced the coroner system with the medical examiner system in 1915. In 1918, Charles Norris became the chief medical examiner. He realized the importance of having a toxicology laboratory as a part of his organization and started working on obtaining funds for this purpose. In the meantime, Alexander O. Gettler performed chemical and toxicological investigations at his laboratory at Bellevue Hospital. Gettler subsequently established the toxicologic laboratories at the Office of the Chief Medical Examiner. Over the next 40 years, he conducted a wide variety of studies on the theoretical and practical aspects of forensic toxicology. One of Gettler's greatest contributions was formalizing academic toxicology education in the United States under the auspices of New York University. Many of today's renowned toxicologists trained with Gettler, who is considered the founding father of forensic toxicology in this country.

New developments in forensic toxicology are often stimulated by developments in related sciences. The second quarter of the twentieth century saw a rapid growth of pharmaceutical chemistry, which resulted in a new challenge to toxicologists: the analysis of synthetic drugs. Most of these were synthesized to mimic the actions of the related naturally occurring compounds, so that similar or the same analytical methodologies could be applied. Stas's isolation technique continued to play an important role in the initial stages of extraction from the biological matrix.

Toxic substances, for purposes of toxicologic analysis, are classified according to the methods used to isolate them from biological material, rather than according to their chemical, physiological, or pharmacological characteristics. Based on this classification, the major groups are as follows:

1. Gaseous and volatile substances (isolation by diffusion or distillation)
2. Metals (isolation by oxidation of organic matter)
3. Toxic anions (isolation by dialysis or ion-exchange methods)
4. Nonvolatile organic substances (isolation by liquid–liquid or liquid–solid extraction)
   a. Organic acids (extractable from aqueous acidic medium)
   b. Organic bases (extractable from aqueous basic medium)
   c. Neutrals (extractable from either acidic or basic aqueous medium)
   d. Amphoteric substances (extractable at their isoelectric point)
   e. Quaternary ammonium compounds (water soluble, extractable as ion pairs with appropriate complexing agents)
5. Miscellaneous (special extraction techniques required)

This is a general classification; many substances can be isolated by more than one of the methods just named and some may require a combination of methods for efficient extraction.

Modern Toxicology

The advent of modern instrumentation, in addition to the continual broadening of the spectrum of available drugs, began changing the classical methodologies in forensic toxicology in the late 1940s and early 1950s. Paper chromatography, developed by Tswett in 1906, began to be replaced by thin-layer chromatography. Ultraviolet, visible, and infrared spectrophotometry became important tools in the identification of isolated substances. Gas chromatography (GC) and high-pressure liquid chromatography (HPLC) followed, and it wasn't long before GC coupled with mass spectrometry (GC–MS) became a "must" as a confirmation technique. The development of immunoassays enabled toxicologists to perform direct tests for certain drugs and avoid the tedious extraction procedures. The application of HPLC–MS, MS–MS, and supercritical fluid chromatography is now being incorporated into many laboratories. The continual improvement of available methodologies also increases the sensitivity of testing and affects the classical approaches to analysis. There is no longer a need to extract hundreds of grams of tissue to confirm the presence of toxic agents, which makes it possible to use postmortem specimens that are available in limited amounts, such as vitreous humor, or to analyze specimens containing low concentrations of drugs, such as hair. And, of course, computer applications have become routine in toxicology, just as they have in other sciences.

Last but not least, the opportunities for formal education and for practical training in forensic toxicology have greatly expanded in the last few decades. The literature is abundant, and in addition to publications in professional journals, many general textbooks are available. Until the mid-1950s, forensic toxicology was usually a major topic included in textbooks on legal medicine.[19–21] Publications that followed[22–32] clearly demonstrate that forensic toxicology has developed into a science in its own right, though it is an integrative science that draws on a knowledge of pharmacodynamics, pharmacotherapeutics, medicinal chemistry, analytical chemistry, pathology, physiology, behavioral sciences, jurisprudence, and other sciences.

The formation of forensic science societies also played a significant role in shaping the future of forensic toxicology. In January 1948, R. B. H. Gradwohl called a meeting at the Police Academy of St. Louis in Missouri to discuss various aspects of forensic medicine. This meeting, attended by about 150 people, is considered the birth of the American Academy of Forensic Sciences. One of the three original sections of the Academy was the Toxicology Section. Over the years, many other professional societies have been organized, culminating in the establishment of the American Board of Forensic Toxicology, Inc. in 1975. The Board at that time defined forensic toxicology as "the study and practice of the application of toxicology to the purposes of the law". This rather broad definition primarily applied to postmortem forensic toxicology. Changes in the legal and government regulatory arenas over the last 10–15 years have resulted recently in the emergence of three defined areas of forensic toxicology, namely postmortem forensic toxicology, human performance forensic toxicology, and forensic urine drug testing.

Murder by poison is relatively rare in today's society, but postmortem forensic toxicology remains an important and integral part of the medicolegal investigation of deaths by medical examiners and coroners. Many unexpected, sudden, or violent deaths are associated with drugs, and the final diagnosis of the cause or manner of death frequently depends on toxicologic findings in postmortem specimens. There are essentially three types of situations requiring answers from the toxicology laboratory: when the cause of death is known but drug findings are needed to clarify the circumstances, when drugs are the direct cause of death, and when negative toxicologic data permit the pathologist, as they often do, to conclude that death was caused by disease.

Clearly, postmortem forensic toxicology testing, just like any area of analytical toxicology, has to yield unequivocal analytical results, a fact that has been recognized since the beginning of the science. Many aspects of the laboratory practice of postmortem forensic toxicology are shared with related areas of toxicology, such as complete, up-to-date procedure manuals, qualified laboratory personnel, security, and chain-of-custody procedures based on accepted standards for forensic specimens, laboratory safety manuals, and the like. The standards for qualitative and quantitative analysis are just as rigorous as they are for clinical toxicology or forensic urine drug testing, and so are the applications of quality assurance and quality control procedures.

Although there are many similarities among the three areas of forensic toxicology, there are also significant differences, which explains why postmortem forensic toxicology, the oldest discipline of toxicology, has been the last to start developing regulatory procedures for its practice. The complexity of postmortem toxicology testing and the nature of specimens submitted for analyses make this area of forensic toxicology unique and difficult to regulate without allowing a certain degree of flexibility. Technology has changed over the years, but some basic principles in postmortem toxicology are the same as those familiar to Orfila or Stas.

The initial systematic approach to postmortem forensic toxicologic analyses can vary from case to case. A certain number of drugs frequently are encountered in the laboratory and therefore included in routine screening procedures. The same initial routine screen may not be suitable for every laboratory. The frequency of occurrence of certain drugs varies with the geographical area. The available equipment and personnel (two factors affected by the laboratory budget) will also influence the selection of methodologies for both qualitative and quantitative analysis. The number of possible analytes in postmortem forensic toxicology is unlimited, and therefore the initial screening can never be all-inclusive. The types of specimens submitted for analysis are not unlimited, but the quality of the biological matrix varies greatly with the degree of decomposition, as does the amount of the specimens available for analysis. The types and minimum quantity of biological specimens required for toxicologic analysis is determined by the analytes, which must be identified and quantified. On rare occasions, the analysis of a single substance in a single specimen will yield a satisfactory answer. Multiple drug ingestion is much more common.

In the ideal situation, it is suggested that the following specimens be collected at the time of autopsy: heart blood (25 mL), peripheral blood (10 mL), liver (100 g), brain (100 g), kidney (50 g), bile (all available), urine (all available), vitreous humor (all available), and gastric contents (all available). Collection of all these specimens does not necessarily mean that all of them will be analyzed. Some are more suitable for screening, and in some cases concentrations need to be determined in more than one specimen in order to provide meaningful interpretation of results. Most cases have to be addressed on an individual basis. Unique drugs or chemicals may require special specimens, such as lung, intestine, or synovial fluid. On the other hand, special specimens may be the only ones available, such as muscle, skin, bone, hair, embalmed or exhumed tissue, or even maggots. This variability of postmortem forensic specimens makes the preparation of control materials not only difficult, but in certain instances impossible.

The case history is of paramount importance to postmortem forensic toxicologic investigation because the information provided by the pathologist or investigative officer will to a great extent influence the initial analytical approach and help direct the analyst toward identification of specific substances of interest (in addition to those included in the routine screening procedure). There are instances, however, in which the case investigation and an extensive screening fail to provide any pertinent information regarding the possible toxic agent. In these cases, the analyst has to resort to a comprehensive analytical screening referred to as "the general unknown".[33] The general unknown consists of a series of analytical screening procedures, conducted with the purpose of either excluding or confirming the presence of toxic agents. It is a screen in which the exclusion of as many toxic substances as possible can be just as important as the identification of a specific substance. It is the type of analysis that Jean Servais Stas performed at the scientific dawn of forensic toxicology.

References

1. Doyle, A. C. *A Study in Scarlet*; Beeton's Christmas Annual: London, 1887.
2. Thorwald, J. *Science and Secrets of Early Medicine*; Thames & Hudson: London, 1962.
3. Joachim, A.; Ebers, P. *Das älteste Buch über Heilkunde*; George Reimer: Berlin, 1890.
4. Niyogi, S. K. In *Introduction to Forensic Toxicology*; Cravey, R. H.; Baselt, R. C., Eds.; Biomedical: Davis, 1981; pp 7–24.
5. Plenck, J. J. *Elementa Medicinae et Chirurgiae Forensis*; Vienna, 1781; p 36.
6. Orfila, M. J. B. *Traité des poisons tirés des règnes minéral, végétal et animal ou Toxicologie Générale*; Crochard: Paris, 1814.

7. Marsh, J. *London Medical Gazette* **1836**, *18*, pp 650–654.
8. Nash, J. R. *Look for the Woman;* M. Evans and Company: New York, 1981; p 243.
9. Thorwald, J. *The Century of the Detective;* Harcourt, Brace and World: New York, 1965; pp 296–306.
10. Bouchardon, P. *Le Crime du Chateau de Bitremont;* Albin Michel: Paris, 1925.
11. Stas, J. S. *Bull. Acad. Roy. Med. Belg.* **1851**, *11*, pp 304–312.
12. Otto, F. J. *Ann. Chemie.* **1856**, *100*, p 39.
13. Beck, T. R.; Beck, J. B. *Elements of Medical Jurisprudence,* 12th ed.; (revised by C. R. Gilman). Lippincott: Philadelphia, PA, 1863.
14. Wormley, T. G. *Micro-chemistry of Poisons;* Bailliers Brothers: New York, 1867.
15. Reese, J. J. *A Manual of Toxicology;* Lippincott: Philadelphia, PA, 1874.
16. Reese, J. J. *The Text Book of Medical Jurisprudence and Toxicology;* Blakiston: Philadelphia, PA, 1884.
17. Hemming, W. D. *Forensic Medicine and Toxicology;* Putnam: New York, 1889.
18. Witthaus, R. A.; Becker, T. C. *Medical Jurisprudence, Forensic Medicine and Toxicology;* William Wood: New York, 1894; Vols. 1–2; 1896; Vols. 3–4.
19. *A Text-Book of Legal Medicine and Toxicology;* Peterson, F.; Haines, W. S., Eds.; W. B. Saunders: Philadelphia, PA, 1904.
20. Gonzales, T. A.; Vance, M.; Helpern, M.; Umberger, C. J. *Legal Medicine: Pathology and Toxicology,* 2nd ed.; Appleton-Century-Crofts: New York, 1954.
21. *Legal Medicine;* Gradwohl, R. H. B., Ed.; C. V. Mosby: St. Louis, MO, 1954.
22. Kaye, S. *Handbook of Emergency Toxicology: A Guide for the Identification, Diagnosis and Treatment of Poisonings;* Charles C Thomas: Springfield, IL, 1954.
23. *Toxicology;* Stewart, C. P.; Stolman, A., Eds.; Academic: New York and London, 1960; Vols. 1–2.
24. Curry, A. S. *Poison Detection in Human Organs;* Charles C Thomas: Springfield, IL, 1963.
25. *Isolation and Identification of Drugs;* Clarke, E. G. C., Ed.; Pharmaceutical: London, 1969; Vols. 1–2.
26. *Handbook of Analytical Toxicology;* Sunshine, I., Ed.; CRC: Boca Raton, FL, 1969.
27. *Manual of Analytical Toxicology;* Sunshine, I., Ed.; The Chemical Rubber Co.: Cleveland, OH, 1971.
28. Curry, A. S. *Advances in Forensic and Clinical Toxicology;* The Chemical Rubber Co.: Cleveland, OH, 1972.
29. *Toxicology: The Basic Science of Poisons;* Casarett, L. J.; Doull, J., Eds.; Macmillan: New York, 1975.
30. Baselt, R. C. *Disposition of Toxic Drugs and Chemicals in Man,* Biomedical: Canton, CT, 1978; Vols. 1–2.
31. Houts, M., Baselt, R. C., Cravey, R. C. *Courtroom Toxicology,* Matthew Bender: New York, 1981; Vols. 1–6.
32. *Introduction to Forensic Toxicology;* Cravey, R. H.; Baselt, R. C., Eds.; Biochemical: Davis, 1981.
33. Stajic, M. In *Introduction to Forensic Toxicology;* Cravey, R. H.; Baselt, R. C., Eds.; Biomedical: Davis, 1981; pp 169–181.

8

The Search for Arsenic

Robert H. Goldsmith

The effort to identify arsenic in various samples has steadily progressed from simple test tube reactions to highly sophisticated instrumental analysis and is regarded as quite effective. The toxic effects of arsenic have served homicidal purposes for centuries. Not until the Blandy trial of 1752 were chemical tests first used to detect the presence of arsenic. Early precipitation and simple colorimetric tests were followed by more chemically precise and sensitive tests such as the Marsh and the Reinsch tests. Recent tests have included specific titrimetric and colorimetric tests and various instrumental techniques, notably atomic absorption and neutron activation analysis. In this chapter, specific cases illustrate the value and the limitations of various techniques.

Arsenic enjoys a fascinating history and has been an object of interest not only to scientists and medical specialists but also to mystery writers, amateur detectives, and the average citizen. It has been shown to be a nutrient, a drug, and a poison. This chapter focuses on arsenic's role as a toxic material as well as on exactly how scientists are able to detect arsenic. It reviews the search for arsenic that has been going on for centuries and its fascinating cast of characters and crimes.

Arsenic is found widely in nature. It is not found often in water or air, but some is found in seawater and in the air produced in certain industrial areas. It may also be found as a contaminant in water running off land contaminated with arsenic industrial wastes or areas where heavy spraying with arsenical pesticides has been employed. Arsenic sprays can also find their way into the food chain if a residue persists in fruits, vegetables, or animal feed. Certain plants can absorb arsenic from the soil.

Foods including most fruits, vegetables, cereals, meats, and dairy products that form part of the human diet contain less than 0.5 part per million (ppm) of arsenic (1 ppm is equivalent to finding 1 milligram of arsenic in 1 kilogram of tested material); these foods, when fresh, would rarely have more than 1 ppm.[1] However, because cows sometimes graze on plants that are contaminated with arsenic, one may find cows' milk that has levels of arsenic that exceed 1 ppm. Seafood is frequently higher in arsenic than other foods.

It may be surprising that arsenic is found as a normal component of the human body. Although it is widely spread through the various tissues and fluids of the body, great variation has been found in the amounts of arsenic within these tissues. In one study of healthy human tissues by Smith using a modern technique known as neutron activation analysis, the mean arsenic concentrations of moist tissues were reported to be between 0.04 and 0.09 ppm on the dry basis.[2] This variability has been confirmed, and it has also been shown that the amount of arsenic in the skin, the nails, and the hair is much higher than in the other tissues. There is no evidence that arsenic is extensively stored in any internal organ or tissue. Blood concentrations are also variable, and the results seem to be influenced by the technique used to make the determination. About 80% of arsenic in the blood can be found within the red blood cell. So finding arsenic within human tissues is not by itself a sign of poisoning but a predictable event.

Arsenic shows much variation in its chemical properties. It can react with oxygen to form two different compounds known as oxides, and with sulfur to form four different compounds known as sulfides. A closely related group of compounds contains arsenic, oxygen, and a metal ion and are known as arsenates and arsenites. Potential poisoners have not used elemental arsenic but have employed common arsenic compounds such as the arsenic oxides, especially white arsenic, As_2O_3; the arsenic sulfides such as yellow arsenic, As_2S_3; and arsenious acid, arsenous acid, and the salts of these two acids, which also are known as arsenites and arsenates. Because the chemical tests that have been designed for the detection of arsenic frequently pick up the presence of antimony, an element that resembles arsenic and itself has been used as a poison, supplementary tests are conducted to distinguish between them. Some other elements will also interfere with certain tests for the detection of arsenic.

Arsenic as a Drug and Nutrient

In our examination of the three roles of arsenic as drug, nutrient, and poison, the initial focus is on its role as a drug. A brief glance at its nutrient role follows, and then attention is concentrated upon its role as a poison and the detection of arsenic as a suspected toxic material.

It is believed that arsenic was clearly known about 400 B.C. Hippocrates was known to have given orpiment, As_2S_3, as a remedy for ulcers and other problems, and orpiment was described in the first century A.D. as a useful remedial agent. Arsenic compounds were suggested for use as a sudorific in the fourteenth century by Jean de Gorris, an antiplague agent by Salva, an antimalarial material by Lentilius (1684) and Friceius (1710) and as an amulet by Donzellus in 1686 to keep the plague from infecting a potential victim.[3] The word *arsenic* did not appear in the English language until 1389. Various arsenical preparations were found in the seventeenth century as medicinal materials for the treatment of ulcers and cancer (Frere Come's paste for both conditions, Lanfranc's collyrium for ulcers, and Hellmund's ointment as a curative agent for cancer) as well as for numerous other conditions and afflictions.[4]

During the early part of the eighteenth century, physicians used arsenic in both internal and external preparations.[5] One could find arsenical compounds used for antiseptics, antispasmodics, sedatives, tonics, or other reasons, and arsenical preparations with names like Aiken's tonic pills, Sulphur compound lozenges, and Donovan's solution were in active use for the next century and a half. One of the most popular remedies during this period was Fowler's solution, introduced by Thomas Fowler in 1786. It was composed of potassium arsenite flavored lightly with oil of lavender and tincture of cinnamon.[6] It became popular in the treatment of remitting fevers and periodic headaches and as a tonic for the skin, nerves, and blood and for general health and well-being. It became fashionable for the general public to take arsenical preparations as self-administered supplements for their complexion, appearance, and overall good health. In 1910, Paul Ehrlich introduced arsphenamine and related arsenical compounds as effective antibacterial drugs, which became the most important use of arsenic compounds in medicine in this century.

Arsenic has also been classified as an essential mineral nutrient. In 1976, the evidence for essentiality became conclusive. Arsenic deficiency was characterized by depressed growth and abnormal reproductive patterns in several different species.[7] It has been suggested that arsenic is involved in the metabolism of the essential amino acid methionine. While no recommended daily allowance (RDA) has yet been established for arsenic, animal studies suggest that the dietary needs are in the range of 6.35 to 12.5 micrograms (μg) per 1000 kilocalories of energy consumed in food and drink. It has been previously pointed out that fish and seafood in general are among the best sources of the element.

Arsenic as Poison

The toxic nature of arsenic has probably been known since pre-Christian times. History has shown arsenic as the instrument for many deaths. The Borgia family of the fifteenth and sixteenth centuries were practitioners of the art of poison with arsenic. About 600 murders during the 17th century were attributed to Tofana, who devised arsenical powders, a variety of arsenic-containing pills, and various solutions with arsenical materials.[8] In the sixteenth century, Catherine de Medici is alleged to have tried to dispose of Henry of Navarre using arsenious acid solutions placed on the leaves of a book and allowed to dry so that as an unsuspecting Henry repeatedly touched his finger to his mouth and then the page, he would be taking the poison gradually.[9]

The nineteenth century had its share of arsenic poisoning cases. Of 541 fatal poisoning cases in England and Wales during 1837 and 1838, 185 were connected to arsenic.[10] Data from France for 1832–1840 connected 141 out of a total of 194 poisoning deaths with arsenic. Although the number of deaths linked to arsenic has declined dramatically in the twentieth century, there seems to be no reason to remove it from our list of potential poisons, despite the greater difficulty in obtaining it and the greater ability of science to detect it. A study conducted in North Carolina pointed out that 11 homicidal poisonings done with arsenic were reported in 1969–1971 and that from 1972 to 1982 a total of 28 arsenic poisoning cases were reported, of which 14 were homicides.[11] For 1978–1980, Duke Medical Center documented an additional 36 cases of hospitalization due to arsenic poisoning.[12] The threat of poisoning with arsenic is still very real.

Levels of toxicity for arsenic have created much confusion. In the mid-1800s, it was publicized that male and female Syrian peasants were regularly consuming arsenic to produce positive effects on appetite, complexion, energy, and general health. Some people also began to eat arsenic to become immune to arsenic poisoning. Many of these reports were disputed, but arsenic preparations were used by many in the general population for a variety of reasons. It is now known that one may consume very small amounts of arsenic without difficulty, and in fact some arsenic is essential. The poison prevention claim is controversial, however, because it has never been proven that tolerance to arsenic can be guaranteed. One difficulty is that the amount of arsenic that produces toxicity varies greatly from person to person. Arsenic eaters who ingest 7 mg of arsenic per day may be taking the less toxic pentavalent form, but about 30 mg of the more toxic As_2O_3 has been shown to be fatal despite the fact that higher levels have been reported as nonfatal in the literature. It does appear that a limited tolerance may exist in certain situations. The levels of arsenic and the chemical form seem to be the most important factors in determining the toxic effects of arsenic ingestion. Very small amounts may produce some biochemical changes, but there is no proof that they will be protective, and unless the dosage and chemical form are known the overall impact cannot be assessed properly.

Testing for Arsenic Begins

The first arsenic case in which chemistry played a role was that of Mary Blandy in England in 1752. Blandy was accused and later convicted of administering white arsenic (arsenous oxide, with a formula As_2O_3) to her father, who was an important solicitor. She had obtained some of the poison through her accomplice, a Captain Cranstoun, who desired to marry her against her father's wishes. The first attempt at poisoning using tea did not succeed, and Cranstoun advised Blandy to use a different medium in which the poison would not float on top of the water. She then used gruel, and not only did her father fall ill but also a charwoman and a maid, arousing suspicion.

An inspection of the pan used to prepare the gruel revealed a white powder at the bottom of the pan. Anthony Addington, medical examiner and chief witness for the crown, in his chemical examination of the substance noted its physical and chemical properties. When the powder was placed in water, it turned milky in appearance. Part of it floated on top of the water like a thin film, but most of the material sank to the bottom. A gritty and insipid taste was noted, and the odor of garlic was clearly evident upon placing it in a red-hot pan.[13] He also found that its behavior was identical to that of the white arsenic sample he tested simultaneously.

Joseph Black in his chemical testing mentioned the odor of garlic, the reduction of the powder to elemental arsenic after heating it with black flux (recently ignited charcoal) in a test tube, and the deposition of the arsenic on a bright copper plate.[14] Chemical equations for the latter two reactions are as follows:

$$2As_2O_3 + 6C(s) \rightarrow As_4(s) + 6CO(g)$$

$$4Cu(s) + 4As^{2+}(aq) \rightarrow As_4(s) + 4Cu^{2+}$$

Chemical testing clearly played a role in the trial of Mary Blandy, in contrast to the case of George Wythe, who most likely was murdered by his grandnephew George Wythe Sweeney.[15] George Wythe was a well-known Virginia patriot who not only was a signer of the Declaration of Independence but had distinguished himself as a lawyer, legislator, and champion of the new republic. At the age of 80, Wythe was living with his sister's grandson, and his legacy was to go to his grandnephew if Wythe died before him, as the grandnephew knew. On May 25, 1806, George Wythe and two of his servants became seriously ill. One of the servants recovered, but the other servant and Wythe died within a few weeks. During this period, the grandnephew was removed from the will.

Sweeney was alleged to have purchased yellow arsenic (arsenous sulfide with a formula As_2S_3) and on the morning of the poisoning one of the servants saw him toss a small white paper packet into the fire. A quantity of yellow arsenic, an easy material to obtain at the time, was later found in his room. Despite much circumstantial evidence Sweeney was found not guilty. For one reason, one of the servants, who had witnessed the most critical events, was not allowed to testify because of her race; for another, the autopsy was not done well. The examining doctors testified to the inflammation of the stomach with the appearance of a black vomit and the presence of other symptoms associated with the administration of yellow arsenic. However, the omission of chemical analysis meant that positive proof of the presence of poison in Wythe's body was never presented, even though in 1806 chemical means to reach a conclusive answer were clearly available.

Early Chemical Tests

Writings from the very beginning of the nineteenth century describe some of the earliest tests employed in the search for arsenic. Joseph Black in 1803 described those qualities believed to distinguish arsenic from other materials, such as its weight; its volatility; the fact that white arsenic treated with black flux would give rise to elemental arsenic; the ability of arsenic upon being heated using a dull red heat to penetrate metallic copper and give the copper a whitish color; and arsenic's quality when burned of yielding either a whitish smoke or, if only evaporated, a garlic odor in its vapors.[16] Thomas Ewell in 1806 argued that the presence of arsenic was demonstrated when one detected white flames and the smell of garlic after throwing a small quantity of the suspected powder upon heater coals.[17] At about the same time, Benjamin Rush described the tests that he used for detecting arsenic as being the presence of the garlic smell, the appearance of a whitish presence on copper plates after heating the white powder between the two copper metallic plates, and the formation of a green precipitate when the whitish powder is treated with alkaline copper sulfate.[18]

The last test mentioned by Rush is an early version of one of the three most common precipitation tests. Also known as Green's test, this precipitation test involves treatment of the suspected arsenic solution with ammoniacal copper sulfate.[19] This solution, when added to solutions of arsenites or arsenates, will produce a green precipitation of copper arsenite or a greenish-blue precipitation of copper arsenate. The equation for this reaction is

$$3Cu^{2+} + 2AsO_4^{-3} \rightarrow Cu_3(AsO_4)_2(s)$$

With organic mixtures, supplemental tests on the precipitate to certify a positive arsenic presence will be necessary. Green's test was considered a sensitive test, but it was pointed out that the time required for the precipitate to form varied according to the concentrations of the respective solutions.

The second precipitation reaction is commonly known as Hume's test. In 1809, Joseph Hume noted that the addition of silver nitrate solution to arsenious acid or an arsenite solution in an alkaline environment would immediately yield the bright yellow silver arsenite, Ag_3AsO_3, precipitate.[20] The equation for this reaction is as follows:

$$3Ag^+(aq) + AsO_3^{3-}(aq) \rightarrow Ag_3AsO_3(s)$$

Less frequently, the addition of silver nitrate solution to arsenates yields the reddish brown precipitate of silver arsenate, Ag_3AsO_4.[21] However, silver nitrate tests have no value in the presence of vegetable or animal matter, which result in an altered color of the precipitate or no precipitate at all.

The third precipitation test was the only one of the three that the famous toxicologist Christison believed was useful in composite animal or vegetable fluids.[22] This technique involves passing a stream of sulfureted hydrogen (hydrogen sulfide) gas into an arsenic solution previously acidified with hydrochloric acid to produce a bright yellow amorphous precipitate of arsenious sulfide. Detections were possible in solutions containing one part of arsenous sulfide, As_2S_3, in 100,000 parts of a particular liquid. Analysts also were aware of the need to confirm arsenic due to the possible confusion of the sulfide of arsenic with sulfides of cadmium, selenium, tin, and antimony. Fresnius and Bobo recommended the heating of the arsenical mixture in an atmosphere of dry carbon dioxide gas to yield elemental arsenic.[23]

Besides these three tests, several others were available. Small amounts of solid arsenic compounds could be reduced with carbonaceous flux, recently ignited charcoal, to the elemental state of arsenic (Figure 1). Alternatively, if the quantity of arsenical material is reasonably large, a much better reducing agent is soda-flux, sodium carbonate with one-eighth of the sample's mass of charcoal.[24] Other early liquid tests included the use of limewater, which could result in a precipitate such as calcium arsenate and the formation of a green precipitate using potassium chromate.

Marsh Test

Chemical testing took a great step forward with the introduction of the Marsh method. James Marsh, born in 1794, devoted his life to chemistry. He served as the practical chemist to the Royal Arsenal at Woolwich for many years, and in December 1829 he became an assistant at the Royal Military Academy, serving in that position until his death in 1846.[25] Marsh received the Gold Medal of the Society of Arts of London for his work on arsenic detection.[26]

According to the Marsh method, pure metallic zinc, Zn, was added to either sulfuric acid or hydrochloric acid, and once any suspected contamination was removed, the suspected arsenic solution was added to the flask so that the arsine gas, AsH_d, was released along with the hydrogen gas (Figure 2). The equations for this series of reactions are as follows:

$$Zn(s) + 2HCl(aq) \rightarrow ZnCl_2(aq) + H_2(g)$$

$$6H_2(g) + As_2O_3(aq) \rightarrow 2AsH_3(g) \text{ and } 3H_2O$$

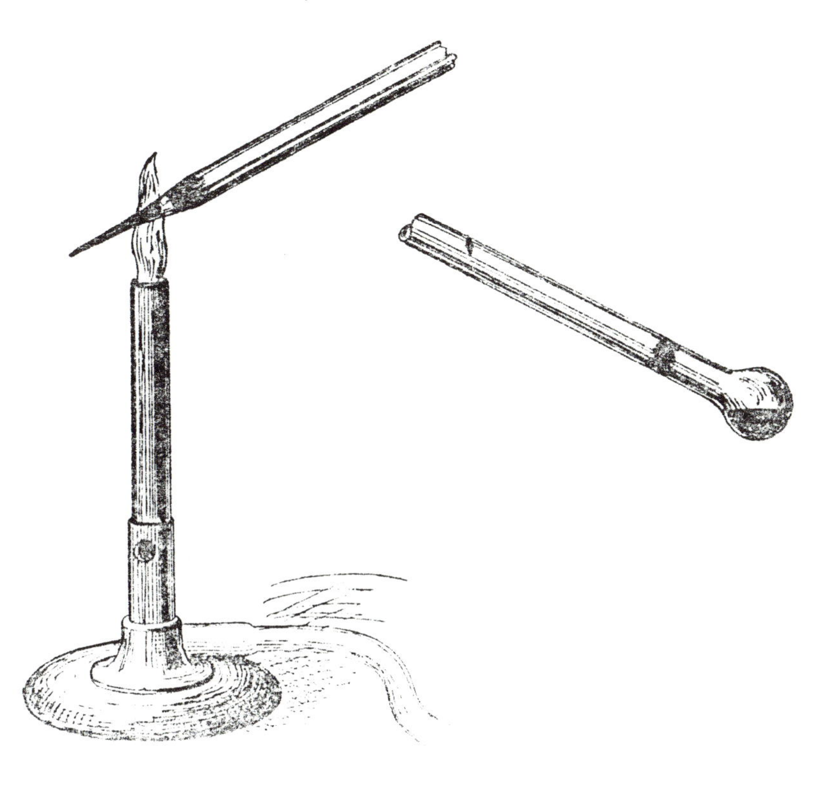

Figure 1. Reduction of an arsenic sample with charcoal. (Reproduced with permission from reference 27. Copyright 1965 The Royal Society of Chemistry.)

Upon heating the gas in a flame, one would find a deposit of metallic arsenic laid down on a white porcelain surface held in the flame or of white arsenic oxide if the dish is held above the flame.

Modifications to the basic procedure appeared regularly. Berzelius modified this test for quantitative analysis by passing the gas mixture through a gas tube heater at the middle so that the arsenic formed at that position could be measured.[27] Other variations in the apparatus have been suggested by Otto and others.[28,29] Methods to clearly identify arsenic in organic matter were developed by Fresnius and Bobo, Gautier, Danger and Flanden, Deflas and Hirsch, Boeke, and others.[30]

Simple confirmatory tests to ensure arsenic rather than antimony was present include arsenic solubility in sodium hypochlorite solution, yellow stain formation by ammonium sulfide, and red color formation for arsenic dissolved in nitric acid and treated with silver nitrate solution.[31–33] Quanti-

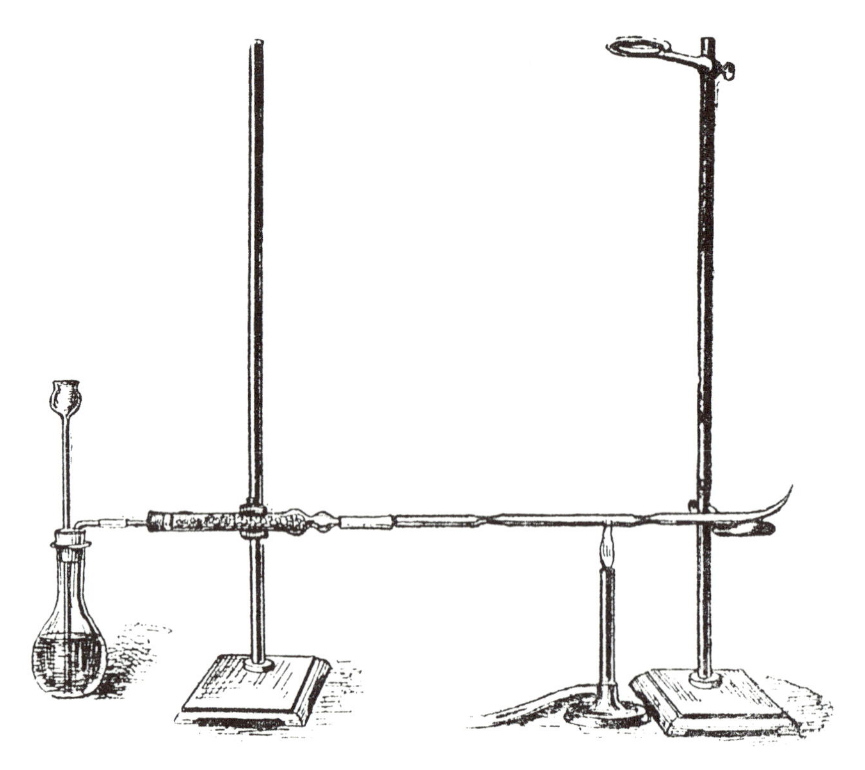

Figure 2. Marsh apparatus. (Reproduced with permission from reference 27. Copyright 1965 The Royal Society of Chemistry.)

tative variations include Evans's measurement of weight increase due to the combination of arsenic from arsine with heated copper to form copper arsenide, Bloxam's method, and the use of electricity to produce measurable hydrogen gas.[34,35]

The Marsh test achieved public recognition in the 1840 case of Marie Lafarge, a young woman charged with poisoning her wealthy industrialist husband with arsenic about a year after their marriage. Mathieu Joseph Bonaventure Orfila, one of the best known and respected toxicologists at the time, originally determined that no trace of arsenic could be found. But due to the sensational nature of the case and the amount of attention devoted to it by the press, the court ordered additional tests on parts of the decomposed body, which had to be exhumed. This time, Orfila detected arsenic in every sample that had been obtained for him from the decomposed body.[36] Marie Lafarge was found guilty and sentenced to life imprisonment. Orfila later commented that a primary reason that arsenic had not been detected in the first place was that a certain amount of experience was necessary to use the Marsh apparatus, which did demand a great deal of practice. The Marsh test gained much publicity and was perceived as requiring a skilled practitioner to obtain meaningful results.

The widespread popularity and use of the Marsh test penetrated fiction in Dorothy Sayers's book *Strong Poison*,[37] featuring the crime-solving aristocrat Lord Peter Wimsey. In this story, Harriet Vane is accused of murdering her former lover, Philip Boyes. Vane is shown to have purchased commercial arsenic, and she even admits to this, stating that as a mystery writer she was trying to show how easy it would be to purchase arsenic. On the day of the suspected murder, the victim had taken lunch with his cousin, the attorney Norman Urquhart, in the presence of Urquhart's maid. Afterward, Boyes went to Vane's apartment, where he had coffee and quarreled with her before he left. Initially, investigators supposed that the poison had been administered in the coffee. The analytical results suggested that 4 to 5 grains (1 grain is approximately 60 mg) of arsenic were administered.

In the course of solving this mystery, Lord Peter and his assistants establish Urquhart's motive, and they locate a mysterious packet with white powder in a secret compartment in the lawyer's office. Lord Peter's manservant, Bunter, who had been taking lessons in Marsh's test, carried out the test on the white powder and saw a deep brownish-black ring with a shining metallic center in the part of the glass tube where the flame impinged upon it. He then dissolved the material in chlorinated lime (calcium hypochlorite) solute to prove that it was arsenic and not antimony. It was eventually determined that the arsenic poison had been inserted through a crack in an egg used to prepare the omelet eaten by both Boyes and Urquhart, with the result that Boyes was poisoned while Urquhart, who had supposedly built up a resistance by consuming arsenic over a period, was not affected.

Emergence of Other Tests

Shortly after the Marsh test came into use, the Reinsch test was introduced as a simple, effective test that could pick up arsenic at a level of approximately 0.00002 part of a solution.[38] The test consists of placing a copper leaf or copper plate or other form of copper metal previously treated with dilute nitric acid into an arsenical solution that had been previously acidified with hydrochloric acid and heated nearly to boiling. Arsenic then reveals itself as a brilliant gray metallic-like coating or a black coat.

The test evoked much commentary after its publication. Ellis thought it bore a close similarity to the Marsh test, and he passed arsine gas over copper to demonstrate the deposition of arsenic on copper.[39] Gardner's experiments suggested that this test was twice as sensitive as Marsh's test.[40] However, Evans showed that the two essentials for the Reinsch test to work successfully were halide ions and a certain amount of hydrogen ions to accompany the copper.[41] The Reinsch test was relatively easy to perform and actually was used in the courtroom shortly after its introduction and before all of its potential limitations were known. These limitations were demonstrated at one trial in 1859 where contradictory results were obtained because Reinsch's test did not seem to work when chlorate ions were present.

The reality of arsenic testing was brought to public attention in 1900 when thousands of beer drinkers in Lancashire became ill. The culprit was found to be arsenic in the beer. This arsenic was tracked back to arsenic-containing sugar, which had been prepared in the sugar refinery using sulfuric acid that itself was contaminated with arsenic. A study was set up to evaluate the existing methods of detecting arsenic, particularly in beer, and it was concluded that the Marsh–Berzelius test was the best test to use even though it required the services of an experienced analyst. The study also indicated the value of the Reinsch test because it could be carried out quite easily and the results were produced rapidly in most cases. However, the consensus of investigators was that the Marsh test was preferable.

A variety of tests besides the Marsh and Reinsch tests were used during the first half of the twentieth century:

- *Bittendorf's test*. This modification of the Reinsch test adds a strong hydrochloric acid solution containing stannous chloride, $SnCl_2$, to an arsenious solution to produce metallic arsenic.[42]
- *Gutzeit test*. Arsine gas produced by the Marsh method is dissolved in concentrated silver nitrate solution to yield a yellow color. Mercurous bromide, Hg_2Br_2, was initially found to be the most sensitive reagent. Filter paper saturated with the reagent produces a given level of color depending on the level of arsine gas, so standardized stains can be prepared for quantitative comparisons.[43]
- *Hefti tests*. Several tests use electrolytic deposition of arsine upon an electrode.[44]
- *Kage modification*. The more sensitive mercuric bromide, $HgBr_2$, can be used as the reagent in the Gutzeit test.[45]

- Several methods using titration by iodine and early colorimetric tests using ammonium molybdate were also available but were not seriously pursued in forensics.[46]

Of these tests, the Gutzeit test is still regarded as a simple and rapid qualitative method of arsenic detection.

Modern Tests for Arsenic

Since the mid-1960s, a number of new instrumental and other approaches have made their appearance, including thin-layer chromatography, coulometry, and a more sensitive colorimetric test that uses the chelation of arsenic with the reagent diethylthiocarbamate to form a complex whose color absorbance could be measured. However, none of these have become important forensic techniques.[47,48] The important Gutzeit test was modernized by Curry, who inserted the framed reagent dried filter paper into a spectrophotometer to take optical readings and compare them to various calibrated standards (see Figure 3).[49] The various forms of this test can easily and rapidly detect urinary arsenic even in low concentrations and are still in use today. Chemical spot tests using a combination of ammonium carbonate, ammonium sulfide, and potassium iodide solutions were suggested for use as a single simple test of the suspected metallic poison.[50] The extent of the usage of spot tests in forensics today is not clear.

In the view of many analysts, however, the two most important new techniques are atomic absorption and neutron activation analysis.

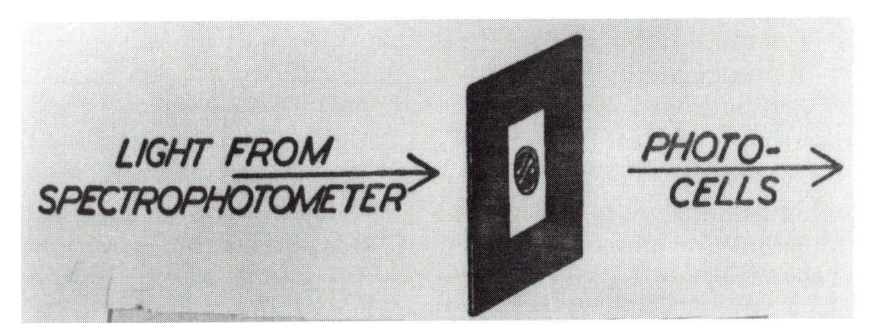

Figure 3. Method of measuring stains from a Gutzeit holder. (Reproduced with permission from reference 49. Copyright 1988 Charles C Thomas, Publisher.)

Atomic Absorption and Neutron Activation Analysis

Atomic absorption is based on the absorption of light by activated atoms (see Figure 4). Electrons move around a normal or unexcited atom in distinct energy orbitals. Certain amounts of energy can be absorbed by this atom so that electrons can move from a low, distinct energy level into a higher energy level known as an excited state of the atom. When the electrons fall back from this unstable, excited state to their original or ground state, a precise amount of energy is released. These energy packets given off by each atom have distinctive wavelengths. Thus, each atom has a distinctive pattern of wavelengths that can be used for identification, much like fingerprints.

Hollow cathode lamps made from the particular metal are used to produce the precise energy emissions of that element. The emissions from the lamp can be absorbed by a volatile sample of the atoms of that same element. The total amount of absorption can be measured and indicates just how much sample was volatilized by passing directly into the flame (flame approach) or heated to incandescence by electrical heat using a furnace (graphite furnace approach). One newer approach is designed to digest the sample, transfer the digest into the arsine generator, and pass the arsine gas into the burner of the machine.[51] This hydride atomic absorption combination method is said to have a detection limit of about 0.05 μg of arsenic compound, compared to the flame approach's detection limit of 0.1 μg/mL and the graphite approach's limit of 0.05 μg per 10 mL of sample.[52] These sensitivities exceeded all previously used chemical methods.

However, an even more sensitive method is neutron activation analysis, regarded as efficient, accurate, and specific (see Figure 5). Neutron activation analysis involves the treatment of the unknown sample with a stream of neutrons. Within the atoms of each sample are the particles that make up the nucleus of the atoms, namely protons and neutrons. Stable atoms have certain ratios of protons to neutrons. If this ratio is upset by the addition of more neutrons to the nucleus, the atom becomes unstable and may emit radioactive energy such as gamma radiation. This gamma radiation can be detected using devices that pick up radioactive emissions. Usually this technique is regarded as very sensitive because it is able to detect levels of elements in the range of 10^{-3} to 10^{-7} μg of the element per gram of the sample. It is also quite specific for a given element, and interference from other elements is kept to a minimum. Sample irradiation is followed by separation of the arsenic as arsine or another arsenic compound and concluded with the radioactive counting.

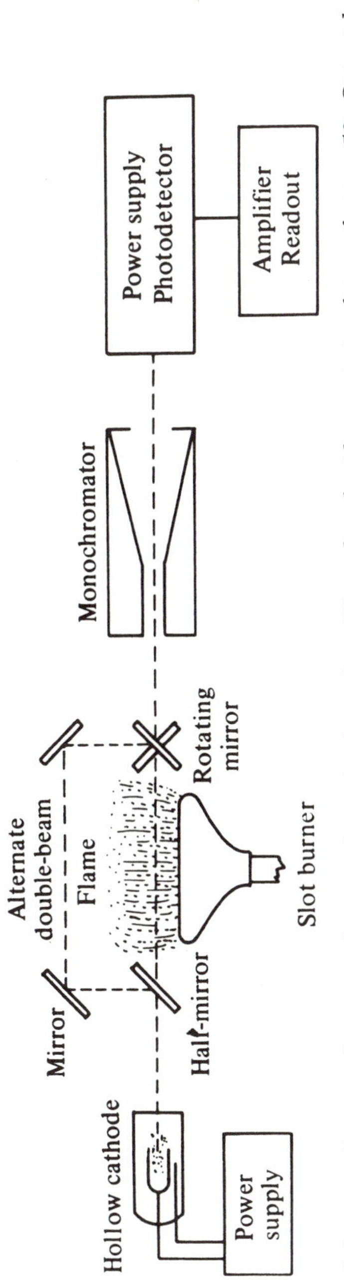

Figure 4. Schematic diagram showing technique of atomic absorption. (Reproduced with permission from reference 59. Copyright 1974 D. Van Nostrand Co.)

Neutron Activation Analysis

In *neutron activation analysis*, a specimen is irradiated with neutrons. This irradiation causes the specimen to become radioactive; that it, it emits gamma-rays. These rays can be measured in an instrument called a *spectrometer*. A *neutron* is an elementary particle that has no electric charge; it is a constituent of all atoms.

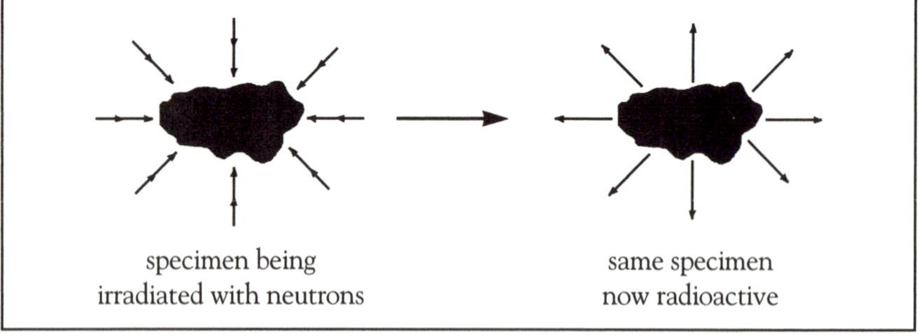

Figure 5. Using neutron activation analysis to detect elements. (Reproduced from reference 60. Copyright 1983 American Chemical Society.)

In the case of *Rex v. Lee*, arsenic levels in the hair and nails of the deceased proved significant.[53] A woman was charged with and later found guilty of using arsenic for homicidal purposes and was sentenced to death for her crime. Besides a medical history that suggested a pattern of the victim's reacting to arsenic ingestion, the amount of arsenic in the stomach, liver, stomach wall, kidneys, hair, and nails was used to prove the charge of arsenic poisoning. Hair lengths of 1/2 in. were taken for analysis. Because hair grows at approximately 1/2 in. a month, the values determined do not show any daily arsenic ingestion but reflect the average intake for a particular month. It now seems possible to correlate the longitudinal arsenic hair levels with a picture of arsenic poisoning. One interesting feature of the case was that the hair sample next to and including the hair bulb had a measured arsenic level of 250 ppm, which was nearly two and one-half times higher than the previous month's level and reflected the final, fatal dose. Similar comments and observations could be made concerning the measurement of arsenic in the victim's nails. Neutron activation analysis has had a clear impact on this process. First, it has made possible the measurement of arsenic in much shorter lengths than were previously employed. Lengths of 1 mm can be used, reflecting intake over a period of only one to two days. Second, this technique has indicated that the normal levels of arsenic in hair are only about 2 ppm, less than half the sensitivity level of previous chemical methods. It should be noted that this technique has also made blood analysis much more sensitive.

New Tests Tackle Old Problems

Neutron activation analysis generated debate when it was applied to samples of Napoleon's hair to determine whether he was poisoned with arsenic. Exiled to St. Helena, Napoleon had generally good health when he first arrived, but his health began to deteriorate shortly thereafter. He exhibited a variety of aches and pains, swollen legs that made it difficult to walk, abnormal sleeping patterns, diarrhea alternating with constipation, headaches, and mood swings. About five years before Napoleon's death, the symptoms became more obvious and severe, and he developed a jaundiced complexion. His health then fluctuated for several years until September 1820. For the next six months, it was generally good but six incidents of severe illness were noted. In his last few months, symptoms became worse, with severe vomiting. Napoleon died May 6, 1821. The original autopsy cited the cause of death as cancer of the stomach, but later data indicated a liver disorder.

A review of the described symptoms suggested that some of them might be due to arsenic poisoning. In the early 1960s, investigators were able to obtain samples of Napoleon's hair, said to have been removed on the day after his death. An arsenic level of 10–38 ppm was obtained using the neutron activation analysis technique.[54] This value, though it includes a wide variation, did suggest much higher than normal levels of arsenic, which could have been due to poisoning, to ingestion of an arsenic medication, or to arsenic-contaminated wallpaper. These same investigators studied arsenic distribution in a 13-cm-long sample of Napoleon's hair and found variable data that suggested intermediate arsenic administration.[55] However, a study 20 years later involving a hair sample owned by the family of a staff officer found normal arsenic levels combined with elevated antimony levels. The report attempted to explain the different results by noting that improvements in the techniques allowed a clear separation of the isotopes of arsenic and antimony. It also suggested that the new study's 12-mg sample would be more representative than the samples of less than 0.5 mg used in the first experiment.[56]

A similar but less well-known speculation occurred in the case of Charles Francis Hall, the American explorer in command of the United States North Pole Expedition, referred to as the Polaris Expedition. He died in his cabin on the U.S. *Polaris* on November 8, 1871. The ship had been anchored off the northwest coast of Greenland only 600 miles from the North Pole. He had suffered with gastrointestinal and other intestinal difficulties since consuming a cup of coffee two weeks earlier. Hall had indicated that he believed someone was trying to poison him, but the inquiry into his death revealed nothing to substantiate this charge. In 1968, a study and autopsy of the body were performed. Samples of hair and nails were analyzed using neutron activation analysis, and the data revealed levels of arsenic that were somewhat higher than normal.[57]

Perhaps the most recent example of a retroactive investigation is the allegation that Zachary Taylor, the twelfth president of the United States, was poisoned by arsenic because of his opposition to permitting slavery to exist in states that would be admitted to the union in the future. The author who made this charge, Clara Rising, was willing to pay a $1200 autopsy fee to have the body exhumed and samples analyzed for arsenic. Samples including hair, nails, and bone scrapings were taken, and arsenic levels were determined using neutron activation analysis as well as other spectrometric methods. These results showed insufficient levels of arsenic to cause poisoning.

Conclusion

Clearly, arsenic has been popular with many poisoners over the years. In small amounts, it has a taste and odor that are not readily noticed. It can be administered in a variety of foods and drink so that it is nearly impossible for the victim to detect it. The fact that it can be given over a long period means that it can produce symptoms that can be identified as a number of other conditions and diseases. But science has developed a variety of techniques that can detect arsenic, provided that an investigator knows the proper method and the limitations of any particular technique. The amounts of arsenic needed for poisoning are detectable in nearly all cases. These minimum toxic levels, as well as the fabled arsenic eaters' capabilities to consume arsenic, were discussed previously.[58]

Today, neutron activation analysis and atomic absorption are the most important techniques for detecting arsenic. Several of the older approaches of wet chemistry are also important. The Gutzeit test and the Reinsch test are equally valuable as the most important qualitative techniques. The Marsh test is used less frequently but still has a number of strong advocates. All other wet tests are far less important.

Looking back, a progression can be seen in the tests, beginning with the purely qualitative and intuitive, moving to crude chemical tests, and now arriving at the more sensitive modern tests for arsenic detection. Arsenic may have a reputation as a popular poison, but it now faces a near certainty of being detected. Therefore, the arsenic poisoner can only be viewed as a person who is willing to take a chance or who is simply unaware of today's successful arsenic searching techniques.

References

1. Schroeder, H. A.; Balassa J. J. *J. Chronic Dis.* **1966,** *19*, 85.
2. Smith, H. J. *Forensic Sci. Soc.* **1967,** *7*, 97.
3. *British Medical J.* **1924,** *1*, 1149.
4. Haller, J. S. *Pharmacy in History* **1975,** *17*, 88.
5. See note 4.
6. Winslow, J. H. *Darwin's Victorian Malady;* American Philosophical Society: Philadelphia, PA, 1971; p 46.
7. Guthrie, H. A. *Introductory Nutrition,* 7th ed.; C. V. Mosby: St. Louis, MO, 1989; p 329.
8. Polson, C. J.; Green, M. A.; Lee, M. R. *Clinical Toxicology,* 3rd ed.; Lippincott & Co.: Philadelphia, PA, 1983; p 421.
9. See note 3, p 1149.
10. Taylor, A. S. *On Poisons;* Churchhill: London, 1848; pp 308–334.
11. Massey, E. W.; Wold, D.; Heyman, A. *Southern Medical J.* **1984,** *77*, 860.
12. *Medical Record Review;* Duke University and Durham Veterans Hospital, 1983.
13. See note 8, pp 440–441.
14. See note 3, p 1149.
15. Boyd, J. P.; Hemphill, W. E. *The Murder of George Wythe;* Institute of Early American History and Culture: Williamsburg, VA, 1955.
16. Black, J. *Lectures on the Elements of Chemistry;* Mundell and Son: Edinburgh, Scotland, 1803; Vol. 2, pp 419–426.
17. Ewell, T. *Plain Discourses on the Elements of Chemistry;* Brisban and Brannan: New York, 1806; p 252.
18. Rush, B. *Medical Inquiries and Observations;* J. Conrad and Co.: Philadelphia, 1805; pp 239–240.
19. Wormley, T. G. *Micro-Chemistry of Poisons,* 2nd ed.; J. B. Lippincott: Philadelphia, PA, 1885; p 263.
20. Hume, J. *Phil. Magazine* **1809**, *33*, 401–402.
21. See note 19, p 261–262.
22. Christison, R. *Edinburgh Medical and Surgical Journal* **1824**, *22*, 78.
23. See note 19, p 268.
24. Christison, R. *Treatise on Poisons*, 4th ed.; E. Barrington and G. Haswell: Philadelphia, PA, 1845; pp 203–205.
25. *Dictionary of National Biography;* Lee, Sidney, Ed.; Smith Elder and Co.: London, 1909; Vol. 12, p 1100.

26. Marsh, J. *Edinburgh New Phil. J.* **1836**, *21*, 229–236.
27. Campbell, W. A. *Chemistry in Britain* **1965**, *1*, 200–201.
28. See note 19, p 280.
29. Gonzales, T.; Vance, M.; Halpern, M. *Legal Medicine and Toxicology*; D. Appleton Century: New York, 1940; p 643.
30. Chittenden, R. H.; Donaldson, H. H. *Am. Chem. J.* **1881**, *2*, p 235.
31. Thompson, L. *Phil. Magazine* **1837**, *10*, 353–355.
32. *Lancet* **1838–9**, *1*, 819.
33. Herold, J. A. *Manual of Legal Medicine*; Lippincott: Philadelphia, PA, 1897; pp 64–65.
34. Evans, B. S. *Analyst* **1920**, *45*, 8–17.
35. See note 19, pp 293–295.
36. See note 27, p 200.
37. Sayers, D. L. *Strong Poison*; Harper and Row: New York, 1938; p 210.
38. Reinsch, H. *Phil. Magazine* **1841**, *19*, 480–483.
39. Ellis, R. *Lancet* **1843–44**, *1*, pp 177–181.
40. Gardner, D. P. *Am. J. of Sciences* **1842**, *44*, 240–249.
41. Evans, B. S. *Analyst* **1923**, *48*, 357–367 and 417–427.
42. See note 19, p 295.
43. Webster, Ralph W. *Legal Medicine and Toxicology*; W. B. Saunders and Co.: Philadelphia, PA, 1930; pp 515–521.
44. Treadwill, E. P.; Hall, W. T. *Analytical Chemistry*, 9th ed.; John Wiley and Sons: New York, 1942; pp 86–87.
45. Gradwohl, R. B. H. *Legal Medicine*; C. V. Mosby: St. Louis, MO, 1954; p 682.
46. Bamford, F. *Poisons*; 3rd ed.; Balkiston: Philadelphia, PA, 1951; pp 84–89.
47. Curry, A. S. *Advances in Forensic and Clinical Toxicology*; CRC: Cleveland, OH, 1972; p 188.
48. Baselt, R. C. *Analytical Procedures for Therapeutic Drug Monitoring and Emergency Toxicology*, 2nd ed.; PSG: Littleton, MA; pp 33–35.
49. Curry, A. S. *Poison Detection in Human Organs*, 4th ed.; Charles C Thomas: Springfield, IL, 1988; pp 103–104.
50. Brookes, V. J. *Poisons*; Robert E. Kreiger: New York, 1975; p 17.
51. *Patty's Industrial Hygiene and Toxicology*, 3rd ed.; Clayton, G. D.; Clayton, F. E., Eds.; Wiley-Interscience: New York, 1981; Vol. 2A, p 1521.
52. *Patty's Industrial Hygiene and Toxicology*; 2nd ed.; Cralley, L. J.; Cralley, L. V., Eds.; Wiley-Interscience: New York, 1979; Vol. 3, p 207.
53. Gordon, I.; Shapiro, H. A. *Forensic Medicine*; Churchill Livingstone: New York, 1982; p 196.
54. Forshufvud, S.; Smith, H.; Wassen, A. *Nature (London)* **1961**, *192*, 103–105.
55. Smith, H.; Forshufvud, S.; Wassen, A. *Nature (London)* **1962**, *194*, 725–726.
56. Lewin, P. K.; Hancock, R. G. V.; Voynovich, P. *Nature (London)* **1982**, *299*, 267–268.
57. Paddock, F.; Loomis, C. C.; Perrons, A. K. *N. Engl. J. Med.* **1970**, *282*, 784–786.
58. Underwood, E. J. *Trace Elements in Human and Animal Nutrition*, 4th ed.; Academic: New York, 1977; p 428.
59. Willard, H.; Merritt, L.; Dean, J. *Instrumental Methods of Analysis*, 5th ed.; D. Van Nostrand Co: New York, 1974.
60. Gerber, S. M. *Chemistry and Crime: From Sherlock Holmes to Today's Courtroom*; American Chemical Society: Washington, DC, 1983; p 69.

9

The Forensic Expert Witness in the Courtroom

History and Development

Jay Siegel

In today's criminal and civil justice systems, expert testimony is an established and accepted part of the process of proof throughout the world. Chemists, biochemists, physicians, engineers, and other experts routinely testify on highly complex and technical matters as diverse as DNA analysis, toxic waste, cause and manner of death, structural integrity of buildings, and the identification of bite marks in the flesh of homicide victims. But expert testimony has a short history—it was not always accepted by courts. Indeed, even the use of nonexpert witnesses is only a few centuries old. This chapter looks back at the development of expertise and expert testimony using the earliest available records.

One of the problems in gathering anything like a comprehensive picture of the use of expert testimony in the past is that keeping records of court proceedings is also a fairly recent phenomenon. Thus, only anecdotal and inferential information is available for ancient times. For this reason, it is instructive to look at the development of a scientific field of expertise in addition to actual examples of testimony; this can provide clues about the kinds of expert testimony that *might* have been offered.

The available literature on the development of expert testimony is sparse. Some sources consider development century by century or milestone by milestone; others discuss the development of expertise by field, such as forensic pathology or questioned documents. The focus of this chapter is how scientific expertise was used in ancient times and how expert testimony developed until, in the sixteenth century, it became an accepted form of evidence in court. Forensic medicine is used to illustrate the use of experts in the early period because it was the first science to have a major impact on adjudication systems. Other scientific fields are also discussed where applicable.

Beginning in the sixteenth century, two major developments occurred in the regulation and forms of presentation of expert testimony—the evolution of the opinion rule and the use of hypothetical questions. We see that expert testimony is merely an exception to the rule against offering personal opinions by witnesses in court. The hypothetical question began as a method to elicit an expert opinion from someone who did not have personal experience in the facts of the case.

Finally, this chapter looks at the central problem with expert testimony—contradictory evidence—and how the jury is to cope with it. Suggestions are offered on how to reform expert testimony in the adversary system.

Evidence and Experts

The unit of currency in criminal and civil litigation is *evidence*. Thayer has defined evidence as "a term of forensic procedure which imports something put forward in a court of justice". He goes on to define the *law of evidence* as "having to do with furnishing to the court of matter of fact, for use in a judicial investigation".[1] Wigmore defines evidence as "any knowable fact or group of facts, not a legal or logical principle, considered with a view to its being offered before a legal tribunal, as to the truth of a proposition, not of law or of logic, on which the determination of the tribunal is to be asked".[2]

Decisions about the role of evidence in a particular case are made by a jury—a group of rational people with an ordinary degree of knowledge and experience.[3] At times, however, people of ordinary knowledge cannot make all of the inferences called for in consideration of particular pieces of evidence. At such times, the trier of fact must be availed of expert testimony by a witness who has the requisite knowledge and experience to make inferences based on technical matters.

Many commentators and legal scholars have offered various definitions of *expert*. For example, Wigmore defines an expert in the context of expert testimony as someone able to assess certain "matters [because they have] some special and peculiar experience more than is the common possession".[4] The New York Court of Appeals defines the expert as "one instructed by experience, and to become one requires a course of previous habit and practice, or of study, so as to be familiar with the subject".[5] Finally, a particularly succinct definition of the expert is given by Rogers: "one who is skilled in any particular art, trade, or profession being possessed of peculiar knowledge concerning the same".[6]

It would be unavailing, if not impossible, to list all of the possible uses of expert testimony. Rosenthal[7] cites several instructive historical examples of cases that show the breadth of the subjects that have required the use of experts. These include medicine, X-rays, electricity and electric lights, chemistry, radio tubes, nautical skill, growth of trees, handwriting, streetcar operation, curative powers of mineral water, fingerprints, ballistics, mental condition, coat alteration, and cellar watertightness.

When most people consider experts and expert testimony, they think of science. They think of experts as giving scientific testimony and of expertise as being knowledge of technical matters involving the sciences. This is not, of course, rigorously true: an auto mechanic can render expert testimony about the condition of a set of brakes, which is a matter of technology, not science. There is by no means a universally accepted definition of *science*, but a few commentators have dared to try coming up with one. One of the more interesting definitions of science, given in the context of the scientific expert in court, has been rendered by Kenney.[8] He lists four criteria for defining when a body of knowledge is a science:

1. *The discipline must be consistent.* Its central principles must be agreed upon by all its adherents.

2. *It must be methodical.* There must be well-defined procedures for collecting data and information about the discipline.

3. *It must be cumulative.* Major principles of the discipline must be able to be carried forth to succeeding generations without being continuously reinvented.

4. *It must be predictive.* It must be able to allow predictions of what is not known from what is already known.

The Concept of the Trial

In the United States and many other countries, the modern notion of a criminal or civil trial involves two sides: the prosecution (or plaintiff in a civil case) and the defense. Each side presents its evidence before the finder of fact (judge or jury), who then must render a decision based on the evidence and applicable laws. In the adversary system, the trial is akin to a battle, and after the smoke clears, the truth is supposed to emerge. Today the central part of the trial process is the presentation of evidence by witnesses who come forward and offer evidence. The form and scope of permissible evidence is governed by the law of evidence and the opinion rule. The general rule is that, with the exception of expert testimony, witnesses must testify to the evidence of their five senses—they are not permitted to draw inferences from the facts. Experts, on the other hand, are needed precisely because they are competent to draw conclusions from facts that the trier of fact is not capable of. This point will be discussed in detail later. It should be emphasized again, however, that this system of presentation of evidence by witnesses, expert and lay, is fairly modern. In the early history of trials, expert witnesses were unknown. In fact, there was no testimony of evidence by any kind of witness!

The earliest form of court trial was not a forum for the presentation of evidence and the discovery of the truth thereby. It was little more than a ritual, a formality for affirming the guilt or innocence of the accused. This is not to say that there were no means for discovering and determining guilt or innocence, just that the function of the trial and judge were more or less to make sure that the formal truth-determining process took place and that the burden of proof was placed on the correct party.

These early trials were generally of four types, depending on the era and the country. Probably the earliest type was *trial by battle*. The two sides of the dispute would quite literally fight a battle to determine who was "right". The theory was not just that the victory would go to the strongest or most powerful, but that fate would intervene on behalf of the virtuous party. A similar form was the *trial by ordeal*. In this process, the accused was subjected to a dangerous or even life-threatening situation. If he or she was virtuous, it was believed that some sort of divine intervention would spare his or her life and thus "prove" innocence. The witch trials of colonial America, in which suspected witches were drowned, are familiar examples of ordeal trials. The *trial by compurgation*, or law wager, involved a formal oath of denial of guilt by the accused. If he then could obtain enough witnesses to swear by oath to his denial of guilt, he would win and be found innocent. Finally, there was the *trial by witness*, which was similar to trial by compurgation: the plaintiff or defendant obtained witnesses to express their belief or concurrence in his side of the argument. In all of these early forms of trial the judge or judges had a limited role; the court acted to ensure that the proper parties were held responsible for establishing the required proof and to make sure that proper procedures were followed.

Gradually, these older types of mechanistic ritual trials began to be replaced by the now-familiar concept of *trial by jury*. This took a long time to evolve, however. Not until the sixteenth century did trials begin to function in the manner with which we are now familiar. Eight hundred years ago, juries were made up of citizens of the locality where the dispute or crime took place—men who were brought together to act as both judges and witnesses. This type of jury would have been aware already of the incident or its members would be expected to find out what they needed to know; the court was not concerned with how the jury was informed and there were no provisions for testimony in court by witnesses. Indeed, voluntary witnesses were frowned upon because of the chance that they could be influenced to alter their story. Not until the sixteenth century in England was an act promulgated that provided a process by which witnesses could be compelled to testify at trial.[9]

Early Examples of Expertise and Expert Testimony

Medicine is one of the oldest sciences and the earliest one with a major impact on legal proceedings. Therefore, it is not surprising that many of the earliest accounts of forensic science in general and the use of experts in particular are found in the area of forensic or legal medicine (also called medical jurisprudence and judicial medicine). Camps defines legal medicine as "the application of medical knowledge to the administration of law and to the furthering of justice and, in addition, the legal relations of the medical man".[10] The first use of such expertise may have occurred about 2500 B.C. in a personal identification based on unusual dental remains consisting of two molars that had been connected by a fine gold wire.

Much evidence analyzed by crime laboratories today, such as weapons and gunshot residue, controlled substances, and much of what is now considered trace evidence, was either not around in ancient times, or the means to analyze it, such as microscopes, were not available. It is interesting to note, however, that the study of handwriting and questioned documents and knowledge of poisons are also quite old, and some early examples of experts on these topics can also be identified.

In the earliest period of forensic medicine there were no specialists and there was no discrete, written body of knowledge. Instead, medically trained people acted as experts as the need arose. Evidence of this is anecdotal, found in ancient sacred books and early law codes. Only at the end of the sixteenth century did a separate field of legal medicine began to appear in a few countries, particularly Italy and Germany. More formal writings and legal proceedings involving the use of medical experts also appeared then, so determining how medicine was used in the courts at the time becomes easier.

In the earliest times, medicine and law were practiced by the priest. Disease and death were considered punishment for transgression of holy laws. Magic and evil spells could also result in disease. Disease was treated by the priest through prayer or even direct intervention; thus the priest can be viewed as the earliest judge and lawmaker and, in addition, the forerunner of the physician.

Medicine and Law in Ancient Times

There is ample evidence of knowledge of disease and of legal codes that regulated the practice of medicine in ancient civilizations. Examples include the Code of Hammurabi (the oldest law code), the Laws of Manu in India, and the laws of ancient Egypt, China, and Persia. Although there are no examples of the use of medical testimony in Greece, Archimedes (287–212 B.C.) was apparently consulted by a king who suspected that his crown was made with an alloy rather than pure gold. This is probably the earliest documented instance of physical evidence analysis. The earliest record of a murder trial has been found in Babylon, recorded on a clay tablet that dates back to about 1850 B.C.[11]

Probably the first medicolegal expert was Imhotep, who lived about 5000 years ago. He was chief physician to the court of King Zozer of Egypt and, in addition, chief justice of the pharaoh's court. Because he held both medical and legal offices, he may be considered the first medicolegal expert. However, no specific records attest to his use of medical knowledge in the courts.

Certainly one of the fathers of medicine was Hippocrates (born 460 B.C.). He and his followers were responsible for great advances in medicine and in the field of jurisprudence. Although there are no written records of the use of medical testimony in court by Hippocrates or his immediate disciples, it is known that he discussed what would now be considered medicolegal questions, such as the effects of wounds on the body and the duration of pregnancy. At about this time the first use of tooth impressions to identify sealing wax for personal identification was demonstrated. For example, Nero's wife was identified by dental work in about A.D. 66.

The Roman era set the stage for the use of medical evidence in the courts, although there are no written records of the use of such evidence in specific trials in early Roman times. It is noteworthy that the body of Julius Caesar was examined by a physician, who declared that only 1 of the 23 stab wounds he received was fatal. The earliest known written example of the requirement of expert medical opinion appears to have been in a Roman document, the *Lex Aquillia*, of 572 B.C. If a slave were to die of negligence after receiving a nonlethal wound, a legal action would have been admissible, but only about the wound, not the death. This means that the severity of the wound was an important consideration in such cases and expert medical opinion would have been required. It is also noteworthy that the XII Tables (449 B.C.) provided for the selection of citizens of Rome to investigate a murder.

Later, during the period of the Emperor Justinian (A.D. 485–565), there were a number of references to legal medicine and the relationship between law and the conduct of physicians. Now part of what is known as the Justinian Code, these references defined an integrated medical profession in detail that had never been attempted before. The function of medically trained men in legal matters was clearly defined, including the principle that the medical expert should not be an ordinary witness but should assist judges by the impartial use of his knowledge. This dictum holds true even today and applies to all expert witnesses, not just physicians. The Justinian Code provided for the cooperation of physicians who specialize in a variety of medicolegal areas, such as rape, poisoning, pregnancy, and the like; some scholars believe that these enactments represent the birth of medical jurisprudence.

Developments After the Fall of Rome

The Roman Empire was overthrown in the fifth century in western Europe by Germanic and Slavic tribes, who introduced the concept of the *wergeld.* This was a monetary payment made by someone who inflicted injury or death upon someone else. The payment was made to the victim or, if the victim died from the injuries, the victim's family. The amount of this payment depended upon the severity of the injury and the status of the person injured. There were also provisions for compensation in the event of injury to animals or property. All of this required that the severity of the injury be determined by the courts. This necessitated the calling of medical experts as witnesses; records of this time make clear reference to the use of such expertise in the courts. During the next 500 years, many laws were drawn up by these tribes. Among other provisions, the *Lex Alemannorum* ordered that "competent men" should examine wounded people and report their findings.

One of the most explicit examples of the use of medical experts occurred during the reign of Charlemagne (742–814). He made great strides in codifying the laws of his empire. These laws included the *Capitularies,* which stressed the need to include all relevant evidence in judicial proceedings. Many examples are found that imply that physicians were often used to provide expert evidence. These include an explicit instruction to judges to incorporate medical expertise, especially in cases of wounds, homicide, suicide, and other crimes.

The Assizes of the Kingdom of Jerusalem were framed in 1100 by the crusader Godfrey de Bouillon. They required that cases of murder be investigated by examination of the corpse (but no autopsies were performed). A report to the court had to be made that included the locations and types of injuries and the likely cause of death.

As with many of humankind's developments, the Chinese were way ahead of the rest of the world in legal medicine. The *Hsi Yuan Lu* (Instructions to Coroners), published about 1250, contains five volumes that deal with many types of death, criminal abortion, inquests, and poisonings. It includes provisions requiring mandatory inquests in suspicious deaths and testimony from corpse examiners.

New Rules in Europe

During the twelfth and thirteenth centuries, Roger II of Sicily and Frederick II of Italy promulgated a number of rules pertaining to the conduct and training of physicians. During this time, Hugo de Lucca, a famous surgeon, took an oath as a medicolegal expert and gave evidence in an abortion case. Many cities in Italy had laws relating to the use of medical expertise.

In 1209, Pope Innocent III declared that physicians should be appointed to the courts to determine the nature and severity of wounds. This arose from the case of a thief who had stolen from a church and had been beaten to death by a number of people, including a priest. A physician was called to the ecclesiastical court to determine if the priest had delivered the mortal blow with a spade. In 1234, Pope Gregory IX had a set of Decretals, the *Commilatio Decretalium,* prepared. These called for the use of medical experts in cases involving torture, a practice that continued into the eighteenth century. Later, Pope Gregory XII reaffirmed the edicts of Gregory IX, making the use of medical expertise obligatory.

Medical experts are known to have been used since the eleventh century in France. Philip the Bold, in 1278, required that surgeons be used in legal matters needing their expertise. In Paris, physicians and surgeons made examinations and reports on people who were injured or died under suspicious circumstances. Medical doctors and midwives were used increasingly by the courts to examine wounds and murder weapons and give expert opinions on matters of poisoning, sexual offenses, and pregnancy.

In fifteenth century Germany, a comprehensive code of laws required medical expertise to be employed in all cases of violent death. The *Carolina Constitution* of 1532 called on the courts to compel medical testimony in cases involving violent or suspicious death. This document put medicolegal practice on firm footing for the first time.

The Law of Evidence and the Opinion Rule

By the sixteenth century, the concept of trial by jury was well developed and accepted. In addition, the concept of proof by testimony of witnesses had become recognized as a proper part of the adjudicative process. These developments gave rise to the law of evidence, which defined what evidence, under what conditions, would be permitted to be heard by the trier of fact. The law of evidence, as developed at that time, was exclusionary in nature and remains so to this day. The original purposes of the exclusionary rules were to protect the jury from misleading testimony and to prevent the introduction of irrelevant and collateral evidence.

The principal exclusionary rule affecting expert testimony is the *opinion rule*. As used today in the courts, the opinion rule generally limits witness testimony to "facts" derived from the experience of the witness's senses. The witness is not permitted to draw conclusions or inferences from these facts; that is the province of the trier of fact. Opinions should only be given in cases where the facts are of such a nature that the trier of fact is not competent to draw needed inferences. Wigmore contends that the theory behind the opinion rule is to keep irrelevant and superfluous testimony out of the proceedings.[12] Many commentators claim, as the rationale for the opinion rule, that a witness who offers conclusions that are within the competence of the trier of fact is usurping its function. Put another way, such opinions concern the ultimate issues before the trier of fact and are therefore not permissible.

Expert Witnesses as Exceptions to the Opinion Rule

One major exception to the opinion rule is that it does not apply to expert witnesses. But the question arises as to how such testimony should get around the tradition that testimony should be based only on personal observation. When experts were embodied in specialized juries or retained by the court as advisers rather than as witnesses, this was not a problem. When experts began to testify regularly in court and offer opinions concerning matters about which they had no personal knowledge, it became necessary to overcome the notion that opinion was not evidence—that only facts were evidence.

This matter was finally settled in 1782 in the case of *Folkes v. Chadd*.[13] The issue in this case was why a harbor filled up with silt, rendering it useless. The plaintiff called a Mr. Smeaton, an engineer, as a witness to give an opinion as to the cause of this occurrence. The defendant objected to this testimony on the basis that the jury should decide the merits of the issue not on opinion, but only on fact. On appeal, Lord Mansfield held the evidence admissible. He stated in part:

> It is objected to that Mr. Smeaton is going to speak, not as to facts, but as to opinion. That opinion is deduced from facts which are not disputed.... Mr. Smeaton understands the construction of harbors, the causes of their destruction, and how remedied. In matters of science no other witnesses can be called. I cannot believe that where the question is, whether a defect arises from a natural or artificial cause, the opinions of men of science are not to be received.... The cause of the decay of the harbor is also a matter of science.... Of this, such men as Mr. Smeaton alone can judge. Therefore we are of opinion that his judgment, formed on the facts, was very proper evidence.[13]

Thus, by the eighteenth century, the opinion rule forbids evidence in the form of opinions as improper, but an exception is made for experts even though they may have no direct knowledge of the facts.

The Hypothetical Question

The next questions that arose about the presentation of expert testimony were procedural. How does one elicit an opinion from an expert who does not have first-hand knowledge of the facts? More importantly, if the expert's testimony relates to the ultimate issue in the case, how is this to be addressed without usurping the jury's function? The answer to these questions lies in the development of the *hypothetical question*. This type of question enables the trier of fact to determine how much weight to put on the expert's opinion by considering the facts that underlie the opinion. If these facts comport with those accepted by the jury, then the opinion will be considered valid.

Although courts vary widely in acceptance or rejection of hypothetical questions, especially with respect to form, this type of question is subject to a well-developed body of law. Essentially, the rules require that the underlying facts and premises be relevant to the case. If so, the content of the question will not usually be subject to challenge and the theory of the question may be of any type that can be supported by the evidence. The question must also contain any facts that are assumed to be true for the purposes of the opinion. This rules out questions that refer to the evidence in the case in general, such as "Based upon the testimony you have heard, what is your opinion…?" Also ruled out are questions that refer to the testimony of a specific witness, such as "If you assume that the plaintiff (or defendant) was truthful in his testimony, what is your opinion about…?"

In his treatise on the development of expert testimony, Rosenthal[14] recites a number of early examples of the use of the hypothetical question, some of which are recounted hereafter. He notes that, although modern law is strict about the rules governing hypothetical questions, the earliest cases of the use of these questions did not indicate that such rules were applied. For example, in the famous 1665 *Witches Trial* in England, a Dr. Browne was called to give an opinion of testimony of "fact" given by other witnesses.[15] There was no hypothetical question or assumption of the prior facts; Browne was only asked for his opinion.

Other early cases cited by Rosenthal include the murder trial of Earl Ferrers, during which the defendant called an expert on insanity. The prosecutor objected to the following question: "Please to inform their lordships whether any, and which of the circumstances which have been proved by the witnesses are symptoms of lunacy".[16] Defendant's attorney then rephrased the question by setting out specific symptoms and asking his opinion on each. This was perhaps a prototype of the hypothetical question.

Another early case was the trial of Lord Melville,[17] who was charged with embezzlement of public moneys. The prosecutor first established how much money had been misappropriated and then sought to show, through the testimony of an accountant, how much the defendant gained from his activity. The defense objected, arguing that this was merely opinion concerning inferences that the jury could conclude on its own. The judges allowed the evidence, reasoning that the question was hypothetical; the accountant was using the facts of the sums of money and then supplying the mathematical result. Essentially the tribunal concluded that if the jury accepted the misappropriation of the sums of money as fact, then they were entitled to consider the conclusion of the "expert" accountant.

By the early 1800s, the hypothetical question had become an accepted method of eliciting expert testimony. In *Beckwith v. Sydebotham*[18] the issue was the seaworthiness of a ship. Surveyors were called to express opinions. The court ruled that even though the surveyors did not have direct knowledge of the facts concerning the ship, their opinions based on these facts were not prejudicial because the other side could also ask them hypothetical questions that assumed facts favorable to their side of the case.

M'Nagten's case (1843) is renowned as a landmark for establishing a definition of insanity. It is also important as an example of the requirement that the specific underlying facts of a hypothetical question be revealed. During the judges' discussions, one of the questions that came up was this:

> Can a medical man conversant with the disease of insanity, who never saw the prisoner previously to the trial, but who was present during the whole trial, and the examination of all the witnesses, be asked his opinion as to the state of the prisoner's mind at the time of the commission of the alleged crime, or his opinion whether the prisoner was conscious at the time of doing the act that he was acting contrary to law, or whether he was laboring under any and what delusions at the time?[19]

One justice observed that such testimony may be irrelevant because the facts the expert uses to base his conclusions on are not revealed. However, he goes on to say that this has been a regular practice and has not been objected to successfully. But the other justices did not go along with this rationale. Lord Chief Justice Tindal concluded that the jury needed to know which facts the witness was relying upon in order to decide whether or not to believe him. Failing that, the opinion should not be admissible. This ruling cemented the idea that the underlying facts need to be stated in the hypothetical question unless they are undisputed.

The most serious objection to hypothetical questions of this sort, where the underlying facts are not explicitly revealed, is that the function of the jury is usurped. Rosenthal cites an opinion of the New York Court of Appeals in *People v. McElvaine*[20] as an excellent example of the court's recognition of this problem. The appeal to this case was based partly on the propriety of the following question put to a physician on the issue of the sanity of the defendant (the physician had attended the whole trial but had not examined the defendant before the trial): "Based upon the whole testimony of the prosecution and the defense... and everything that you have heard sworn to here, now will you answer the question?" The chief judge of the appeals court, Judge Ruger, in ruling that this question was improper, wrote:

> The witness was thus permitted to take into consideration all the evidence in the case... determine the credibility of witnesses, the probability or improbability of their statements.... It cannot be questioned but that the witness was by the question put in the place of the jury, and was allowed to determine upon his own judgment what their verdict ought to be.

The reasoning is that the jury would not be able to determine which facts were used as a basis for the opinion and therefore could not determine if they accepted the truth of these facts and thus what weight they would give the opinion.

Whither Expert Testimony?

Throughout the development of expert testimony, many commentators have argued that this form of evidence is improper because it exposes the jury to opinions about matters that are rightly within the jury's province. Yet it has also been amply shown that there are situations where evidence is so technical that it is beyond the competence of the lay jury to draw the proper inferences from the facts. This is, of course, the raison d'être for expert testimony. But what of the all-too-frequent situation of a fundamental disagreement among experts on both sides as to how the evidence before the jury is to be interpreted? How can a jury agree on a subject when scientists cannot? How are these conflicting principles to be reconciled?

One possible solution has been around for nearly a century. It was ably espoused by Learned Hand in his landmark treatise on expert testimony in the *Harvard Law Review* in 1905.[21] His ideas are worth exploring for they remain as valid today as they were then.

Objections to Expert Testimony

Hand asserted that the present-day practice of calling skilled experts before the jury cannot be justified on the grounds that there were once good reasons for doing so. In fact, there are good reasons for abolishing this practice. As we have seen, the present method is not the only way of using experts in court. The use of jurors skilled or knowledgeable in the major issue and the use of experts called by the court to render advice to the judge about technical matters were the most common ways of introducing expert testimony in early times. It is to the latter practice that Hand wished to turn for the salvation of expert testimony.

Hand claimed that there is no place in court for witness opinions because they "can have no useful bearing on the case, and trenches on the jury's function. It is the jury that should form the opinion, make the conclusion and say truly—*vere dicere*—the fact, not the witness; he merely says what he knows."[22] He asserted that expert testimony is, in fact, no more than a relic of an age when the opinion rule developed. As it matured and the rule excluding opinions of witnesses took form, expert testimony was an exception to the rule because it seemed to be useful.

Having established that the expert witness is an anomaly, Hand was most concerned about the harm caused by this practice. He cited two difficulties that no one would dispute. First, in an adversarial justice system the expert is a "hired gun" of one side of the dispute. Thus a scientist may have great difficulty remaining objective, diminishing the worth of the scientist's evidence. Hand spoke of the natural bias of the expert who represents one side of the dispute and is paid well to do so.

The other difficulty of the use of expert witnesses in the adversary system is that one expert is subject to cross-examination and to contradiction by other experts. Being a witness and not an adviser to the jury, the expert is subject to the same rules and constraints as other witnesses and thus may be examined by the other side. As Hand explained, cross-examination has only two functions: to elicit new facts or to contradict statements made by the witness during direct examination. The latter is itself a form of bringing out new facts that replace the old ones. If a rebuttal witness is called to contradict the first expert, the new witness's testimony is still regarded as bringing out new facts that are in opposition to those of the first witness.

If an expert witness does not have any direct knowledge of facts that are concerned with the main issues in the case, the hypothetical question may be used to elicit a conclusion that will be favorable to the side using the witness. Then this witness is subject to cross-examination and there may be one or more rebuttal witnesses. The result is that two (or more) expert witnesses may disagree as to the inferences that can be made from the same set of facts. If these experts, who are presumably educated and trained to draw inferences from technical matters, cannot agree on the conclusions that should be drawn, how can one expect the jury, whose lack of expertise required experts in the first place, to be able to determine which side to believe? If there is no agreement on the part of the experts, the jury is no better off than if there were no expert testimony to start with.

One Alternative: A Tribunal of Experts

As Hand so ably demonstrated, these are the two fundamental problems with the use of expert testimony in an adversarial system of adjudication. But what is the solution? Courts simply cannot do without experts. The world has gotten too complicated and technical for that. One answer could be for the *court* to call experts to act as advisers rather than as witnesses. This is, of course, not a new idea. As we have seen, some of the earliest uses of experts were in this manner. In fact, many jurisdictions in the United States permit the judge to seek the advice of experts beyond those that may be proffered by either side. In certain countries in South America, the judge has a proactive role in the investigation of crime and may go so far as to tell the prosecutor what types of expertise should be used in proving the case.

Insofar as Hand and others view the use of experts by the court, they envision a sort of standing tribunal of experts in various fields that may be used as advisers to the court. Under one scenario, each side in a litigation or criminal trial would be free to call whatever experts they wish and argue their points in open court in the same manner as at present. At this point, however, the final determination of which point of view should be accepted by the jury would be in the hands of the advisory tribunal of experts. The tribunal would make its findings to the judge.

This "ruling" would then be treated in one of two ways by the court. It could be viewed as an instruction that the court gives the jury, much as the jury may be given instructions on the law. The jury would take the expert tribunal's conclusions as fact. This might cause constitutional problems, however, because it could be interpreted as usurping the jury's function, especially in cases where the verdict hinges on conclusions provided by experts—a situation in increasing abundance these days.

The other way that the conclusions of the court-appointed expert tribunal could be treated is as evidence offered to the jury just like any other evidence. The jury would then be free to give whatever weight it sees fit to the expert conclusions—just as it does now. This seems to put the situation right back at the beginning! But, as Hand wisely counseled, a "reasonable man" test can be used here. This means that although a jury could overrule the decision of the expert tribunal on a given issue, it could only do so when a reasonable man with the same information would overrule. If the jury were to disregard the opinion of the tribunal, the court could take the case away from the jury as it can do now if the jury acts against the evidence. Even if judges were reluctant to take such cases away from the jury, the expert tribunal system would still be beneficial. In cases where jurors get conflicting evidence from the two sides, they would still have the opinion of an impartial tribunal of experts to consider; they would no longer be forced to make uninformed decisions.

Although the court-appointed expert tribunal has several advantages over the present state of affairs in the adversarial system, other problems must be overcome. Chief among these is finding impartial experts. Each state has courts at various governmental levels, and the federal government has its own court system. All of the states and the federal government would presumably need expert tribunals of their own. This would require a good many experts. In the United States, at least, there are a limited number of forensically trained and experienced scientists available. Most of these are employed full-time in government forensic science laboratories; they spend much of their time testifying on behalf of the government in criminal and civil trials and could hardly be candidates for an impartial expert tribunal. This leaves forensic scientists who are in private practice or who teach at universities or colleges. These are relatively few in number and could not possibly supply all of the expert tribunals needed. Another source could be nonforensic scientists who are experts in various fields, such as chemistry or civil engineering, and willing to act as members of an impartial expert tribunal. They would need some education to make them understand how their science fits into the justice system.

Another potential problem with impartial expert tribunals would be the manner of selection of the members. National and regional professional scientific organizations could supply lists of members who are qualified to serve on such tribunals. These organizations could be responsible for the actual selection or could submit nominees to the courts who would make selections. In any case, there must be general agreement in the forensic scientific community that the members of the expert tribunals are both qualified and impartial.

Other Proposals To Improve the Use of Expert Witnesses

Kenney's essay[23] on the expert in court was mentioned earlier in this chapter because it set up useful criteria for determining what disciplines ought to be considered sciences. In addition, he describes what he sees are problems with expert testimony and proposes solutions. He points out correctly that sometimes it is left to the individual combatants in the courtroom to determine if a discipline is a science or not, but the courtroom is clearly not the best place to make that determination and the adversary system is not the best way. Although the context of Kenney's essay is psychiatry and determining whether it is a science (he ultimately decides it does not fit his criteria), the problems he points out are more widely applicable.

Kenney argues that experts too often usurp the function of the jury, a refrain heard frequently throughout the history of expert testimony. He is also concerned about the possible bias that may afflict an expert in the adversary system. Anyone who has testified as an expert in court is aware of the undercurrent of belief that one is either part of the solution or part of the problem, depending on which side one is being used by. Kenney's solutions to these problems are threefold (he is speaking of the English jurisprudence system in his essay, but the same considerations hold elsewhere, especially in the United States):

1. *Statutes that encourage or force experts to render opinions that take them out of their disciplines of science should be abolished.* Here he speaks specifically of the English Homicide Act of 1957.

2. *The courts should not decide what disciplines are sciences.* A registry should be set up that lists those disciplines that are considered sciences. Scientists should decide what new disciplines gain entry. The present system is too ad hoc, hurried, and, well... unscientific.

3. *Expert testimony should be taken out of the adversary system.* The court rather than the parties should call the experts, and they should testify for the court rather than for one side or the other. In addition, the court should be obliged to appoint at least two experts in a given field to testify and the evidence should be admitted only if they agree with each other.

One of the most comprehensive recent discussions of the role of the expert is found in a book by Freckleton, which studies the issues surrounding experts and expert testimony from the Australian experience.[24] Much there can be extrapolated to the United States. He presents a number of problems with using experts in the adversary system and discusses the pros and cons of some suggested solutions, several of which have already been discussed. A few of his more important conclusions include the following:

- *The court-appointed expert.* Freckleton sounds a loud alarm about court-appointed experts. He is concerned that a jury may give a disproportionate amount of weight to the testimony of an "impartial" expert appointed by the court than it would to "partial" experts retained by the parties. Some have proposed that if the court is going to appoint experts, then the parties should be limited in the number that they could retain.
- *Judicial questioning of experts.* Certainly the judge has the right to question witnesses as long as it is not intrusive upon the responsibilities of the jury or counsel.
- *Special juries.* As has already been discussed, one of the earliest forms of expert evidence was impaneling a jury of people knowledgeable in the issues of the case. Although this practice has been abolished in many countries, there have been suggestions that it be revived in certain cases on an experimental basis in England and Australia.
- *Group expert evidence.* This is a variation of the court-appointed expert. Essentially, a group of experts is called by the court to give testimony and answer questions of both parties and the court. In the cases where this has occurred, the testimony was given in a less formal atmosphere and was more of a group discussion than testimony.
- *Availability of forensic scientists and facilities.* Very often, the defense has a difficult time in retaining its own experts and especially in gaining access to facilities for reexamining evidence. In Australia, for example, the independent Forensic Science Centre has been set up to provide equal access to all parties in a dispute.
- *Pretrial discovery of reports.* Countries vary widely in their rules of discovery. Even states within the United States have different standards. One proposal for improving expert testimony is to require complete disclosure of an expert's testimony before trial by dispensing any reports or even the testimony itself as a condition of admitting the expert's evidence. Florida, for example, permits the taking of depositions of all witnesses, including experts, in criminal as well as civil cases.

All the solutions proposed in Freckleton's book have merits and some combination of them may be useful under proper circumstances. Their implementation would surely have wide-reaching effects on the way expert testimony is practiced today.

Where does all of this leave the status of expert testimony? Courts have been using experts in one form or another for over a thousand years. During the last three hundred years or so, the use of expert witnesses has been a commonly accepted exception to the opinion rule. As science and technology (such as DNA typing) provide more critical and pervasive evidence in criminal and civil cases, courts and juries demand more informative and sophisticated experts in court. The stakes have increased and this has spawned more "battles of the experts". The result has been that juries are getting more technical and scientific information to use in helping reach a verdict, but there are also increasing instances of disagreement among experts and confusion among juries. This calls into question whether more expert testimony is really helping the justice system. There is little question that some reforms are going to have to be made in the delivery of forensic evidence. It is equally obvious, however, that change will come slowly and nonuniformly, especially in the United States with its 50 independent state court systems and one federal system.

The Demise of the *Frye* Rule

One of the overarching issues concerning scientific evidence is the conditions under which it is admissible. In many countries, scientific evidence is treated in much the same way as any other type of evidence. Evidence in general is admissible if it is relevant (material and probative) and competent. At first blush, this seems to be a reasonable set of standards for admission; however, there is a danger that scientific evidence may take on an aura of infallibility owing to its sponsorship by a scientist (sometimes called the white coat syndrome). If a well-qualified, eminent scientist offers testimony about a relevant matter, the jury may be tempted to believe it without applying the scrutiny or skepticism that may be applied to other evidence. For this reason and perhaps others, the federal courts in the United States decided in 1923 to adopt a more rigorous standard for the admission of scientific evidence than for evidence in general.

The case chosen to derive the standard was *Frye v. United States*, 54 App. D.C. 46; 293 F. 1013 (1923). This homicide case involved the issue of admission of the results of a test administered to the accused. The test was a forerunner of today's polygraph. *Frye* sought to have the results admitted but was denied by the trial court. On appeal, the appeals court issued the now-famous ruling that not only has served to keep the polygraph out of most court procedures but also has been used to scrutinize many types of novel scientific evidence, including voice print spectrography and DNA fingerprinting. The court's ruling in *Frye* was as follows:

> Just when a scientific principle or discovery crosses the line between the experimental and the demonstrable stages is difficult to define. Somewhere in this twilight zone the evidential force of the principle must be recognized, and while courts go a long way in admitting expert testimony deduced from a well-recognized scientific principle or discovery, *the thing from which the deduction is made must be sufficiently established to have gained general acceptance in the particular field in which it belongs.* [italics added]

Since the *Frye* case, many state and federal courts have wrestled with the concept of *general acceptance*. Some of the evidence of general acceptance includes peer review and publication, evidence of repeatability, and court testimony by experts. The criterion of court testimony by experts has also been addressed in at least one state. In *People v .Young*, 418 Mich. 1; 340 N.W. 2d 805 (1983), the Michigan Supreme Court required that expert testimony offered to support scientific evidence challenged under the general acceptance rule must be given by "impartial and disinterested experts of the relevant scientific community". This means that any expert who derives a substantial part of his other income from the technique is not qualified to offer such testimony.

In June 1993, the U.S. Supreme Court overruled *Frye* in *Daubert v. Merrell Dow Pharmaceuticals, Inc.*, 113 S. Ct. 2786 (1993). This was a toxic torts case that involved testimony concerning the effects of the drug Bendectin. The Supreme Court decided that the *Frye* test has been superseded by federal rules of evidence, especially rule 702, which states: "If scientific, technical, or other specialized knowledge will assist the trier of fact to understand the evidence or to determine a fact in issue, a witness qualified as an expert by knowledge, skill, experience, training or education may testify thereto in the form of an opinion or otherwise."

Under the federal rules and their interpretation by the Supreme Court, scientific evidence is admissible if it is "relevant" and "reliable". Scientific evidence is relevant if it will aid the trier of fact in resolving a factual dispute in the case. Scientific reliability is determined by the degree to which the inference, assertion, or opinion is derived from the scientific method. This evidentiary reliability means that the evidence is scientifically valid; scientific validity means that the underlying theory is grounded in the methods and procedures of science. The Supreme Court rejected general acceptance as the sole criterion to determine scientific validity, calling it too narrow a test, although it is included as one measure of validity. The court suggested three other criteria for determining scientific validity but cautioned that they are not exhaustive:

- The principle has been tested successfully, and the test is repeatable.
- It has been subjected to peer review and accepted.
- Its error rates are known.

In its decision in *Daubert*, the Supreme Court envisioned the trial judge as gatekeeper. The judge has the responsibility for determining if the tests in *Daubert* are met and thus if the evidence should be admitted. The court also relies heavily upon the adversary system to ferret out "bad" science and ensure that the system will work as it is supposed to. This assumption may be unduly optimistic, however, when applied to many criminal cases. It is rare in criminal trials to have scientific experts on both sides; generally, scientific expertise resides in the crime laboratory scientists who have evaluated the evidence at the request of the prosecutor. A lack of resources may prevent the defense from retaining its own experts, and thus the only semblance of an adversary system is the cross-examination of the government's experts. Given that many defense attorneys are unprepared in the theories and practices of scientific analysis, such examinations of experts are often cursory at best; when this is so, no real adversary process exists.

A probable consequence of *Daubert* is that it will be easier to admit scientific evidence than under *Frye*. What is less clear is how the gatekeeper role of the court and *Daubert* will interact to keep out so-called junk science. Some fear that, under *Daubert*, practically everything will come in and the trier of fact will have to sort it all out. Others are more optimistic that the process will be little changed under *Daubert*.

References

1. Thayer, J. B. A *Preliminary Treatise on Evidence at the Common Law;* Little, Brown: Boston, 1898; p 264.
2. Wigmore, J. H. A *Treatise on the Anglo-American System of Evidence in Trials at Common Law,* 2nd ed.; Little, Brown: Boston, 1923, §1.
3. Rosenthal, L. L. *The Development of the Use of Expert Testimony in Law and Contemporary Problems,* Vol. 3; Duke University School of Law: Durhan, NC, 1935; p 493.
4. See note 2, §556.
5. *Nelson v. Sun Mutual Insurance Co.*, 71 N.Y. 453, 460 (1877).

6. Rogers, H. W. *Expert Testimony*, 2nd ed.; Central Law Journal Co.: St. Louis, 1891; p 2.
7. See note 3, p 406.
8. Kenney, D. *Law Quarterly Review* **1983,** 99, 216.
9. Eliz. C.9, §12.
10. *The History of Legal Medicine in Gradwohl's Legal Medicine;* Camps, F. E., Ed.; John Wright: New York, 1976; pp 1–14.
11. Smith, S. *Brit. Med. J. I,* **1951,** *188,* 599–607.
12. See note 2, §§1918, 1920, 1921.
13. *Folkes v. Chadd* (1782) 3 Doug. 157.
14. See note 3, p. 415.
15. *Rex v. Cullender and Duny*, 6 *How. St. Tr.* 1665, 687.
16. *How. St. Tr.* 1760, 886.
17. *How. St. Tr.* 1806, 606.
18. Camp. 116 (1807).
19. Cl. & F. 207 (1843).
20. N.Y. 250, 24 N.E. 465 (1890); see note 3, p. 418.
21. Hand, L. *Harvard Law Review* **1905,** *15,* 44–58.
22. See note 21, p. 44
23. See note 8.
24. Freckleton, I. *The Trial of the Expert: A Study of Expert Evidence and Forensic Experts;* Oxford: New York, 1987.

10

Forensic Science in Detective Fiction

Samuel M. Gerber

Jules Verne, H. G. Wells, and Ray Bradbury are well known. Verne and Wells, particularly, were prognosticators of inventions that came long after they described them. Arthur Conan Doyle, through Sherlock Holmes, described forensic techniques, especially of blood identification, that were not discovered until much later. Equally interesting is the use of forensic science in detective fiction that employs established laboratory methods as well as cutting-edge techniques, such as computer correlation of data and psychological typing. We discuss these authors in this chapter.

"Eureka!" was Archimedes' legendary cry when he discovered a way to determine whether a royal gold crown actually contained adulterating silver. His work is an early example of how physical science may be used to solve a crime.[1]

Ahead of Their Times

Arthur Conan Doyle and Sherlock Holmes

Our discussion of forensic science in detective fiction starts with what is probably the best-known fictional scientific sleuth, Sherlock Holmes.

Doyle[2] wrote 56 short stories and 4 novellas between 1887 and 1927 involving the exploits of Holmes. Science and scientific method are featured in many of them. The generally accepted belief is that Sherlock Holmes was modeled on Joseph Bell, Doyle's professor at the University of Edinburgh, where Doyle received his medical training.[3]

A *Study in Scarlet*, the first story about Sherlock Holmes, was published in 1887. In it, Sherlock Holmes meets Watson for the first time and announces his discovery of a specific test for human blood with the then enormous sensitivity of one part in one million parts of water. He gave few details, of course, because actual specific tests for the characterization of human blood did not exist until much later. The immunological precipitin test was reported in 1901 for identification of human blood. The more general test using benzidine to identify blood from any animal source was announced in 1904.[4]

The use of the microscope for identification of copper and zinc employed by counterfeiters and the linkage of human hair and glue to a specific suspect is described in *The Adventure of Shoscombe Old Place*. The fact that a solution is acid (litmus test) seals the fate of a suspect. The novella *The Sign of Four* describes several forensic techniques. Sherlock Holmes mentions his treatise "Upon the Distinction Between the Ashes of the Various Tobaccos", which he claims enables differentiation of 140 forms of cigar, cigarette, and pipe tobacco. Soil identification and use of footprint casts as forensic tools are also important elements, and Holmes hints at trace analysis with circumstantial identification of a vegetable alkaloid used on a murderous dart.

Several forensic techniques described by Doyle were imaginary or in a rudimentary stage when he wrote about them. Nevertheless, the scientific method and deductive reasoning presaged the development of forensic science.

Umberto Eco and William of Baskerville

Many apocryphal tales involve Sherlock Holmes and similar detectives in other stories. A character strangely reminiscent of Holmes is found in a twelfth-century setting in Umberto Eco's *Name of the Rose*,[5] which describes the investigation by William of Baskerville and his assistant, Adso of Melk, of a series of strange events and a murder in a monastery. (A different Baskerville is featured in Doyle's well-known long tale, *The Hound of the Baskervilles*.) The dust jacket describes William of Baskerville as having "the logic of Aristotle, the theology of Aquinas and the empirical insights of Bacon". In addition, William of Baskerville has many of the physical attributes of Holmes. There are many similarities to the Sherlock Holmes canon in *The Name of the Rose*, which are commented on in *The Key to the Name of the Rose* by Haft, White, and White.[6]

R. Austin Freeman and Dr. John Evelyn Thorndyke

Dr. John Evelyn Thorndyke was the creation of R. Austin Freeman (1862–1943) and the protagonist in 11 novels and 42 short stories.[7] Freeman had a background similar to Doyle's; he was a physician with the same medical specialty (ophthalmology) who turned to writing. Freeman, too, had a role model for Thorndyke: his professor, A. S. Taylor, evidently an early forensic scientist and author of "Principles and Practices of Medical Jurisprudence" (1865).[8] Like Holmes, Thorndyke has a chronicler and co-worker, Christopher Jervis, M.D., as well as a laboratory assistant and general factotum, Nathaniel Polton.

"The Case of Oscar Brodski"[7] involves a brutal murder and robbery. The identity of the murderer is disclosed by a variety of forensic techniques. Thorndyke has an advantage over Holmes: he carries a portable laboratory—"little reagent bottles, tiny test tubes, diminutive spirit lamp, dwarf microscope and assorted instruments". Thorndyke uses his microscope to identify textile fibers and food found on the dead man and links these to items found at the home of the murderer. He uses soil identification to establish the site of the murder. Chemical identification of shellac used to bind rabbit wool in the burnt remnants of the victim's hat is accomplished by extracting the shellac. Thus, in this tale Freeman employs chemical and physical forensic methods in identifying the criminal and reconstructing the crime.

In "A Case of Premeditation",[7] Freeman employs footprint and fingerprint evidence to solve a crime. The use of the Marsh test for arsenic is demonstrated in "The Moabite Cipher".[7] Ballistics involving an air gun is an important feature of "The Aluminum Dagger"[7]; compare it to "The Empty House",[2] of Doyle.

More Ways To Die

Dorothy L. Sayers and Lord Peter Wimsey

Dorothy L. Sayers (1893–1957) is best known for her character, Lord Peter Wimsey, hero of many of her detective novels, but three of her novels and one short story offer good examples of how forensic methods are used to solve crimes.

The Documents in the Case[9] is the most sophisticated story in terms of forensic science. The dead man, a mushroom fancier, is believed to have died as a result of eating *Amanita muscaria* (which is poisonous) that he mistook for *Amanita rubescens* (which is edible). The original verdict was accidental death by muscarine, but it develops that the poisoner had access to synthetic muscarine. Analysis of the residual mushroom stew disclosed that it is a racemic mixture—that is, it contained both the levorotatory and dextrorotatory forms of muscarine, whereas the natural form is completely dextrorotatory. (Dextrorotatory and levorotatory indicate the ability of a chemical solution to rotate a beam of polarized light to the right or left, respectively.) Eh voila!—with other evidence, the poisoner is identified. Natalie Foster gives a detailed explanation in the "Strong Poison: Chemistry in the Works of Dorothy L. Sayers".[10]

Death by arsenic is common in detective fiction. Sayers introduced a new twist in the Wimsey novel, *Strong Poison.*[11] The villain, Urquhart, slowly increases his tolerance to arsenic by taking regularly increased doses. A shared omelet laced with arsenic poisons Philip Boyes with little effect on Urquhart. The intervention of Lord Peter Wimsey lays bare the facts: Urquhart has the anticipated clear complexion and sleek hair, and his nails and hair contain arsenic. Bunter, Lord Peter's man, does the necessary Marsh test for arsenic and even takes the trouble to eliminate the possibility of the presence of antimony.

An unusual type of weapon is the centerpiece of *Unnatural Death.*[12] No cause can initially be found in several deaths, but the cause is ultimately shown to be injection of air with a hypodermic into the unconscious victims. The use of forensic dentistry is offered in Sayers's short story, "In the Teeth of the Evidence".[13] Here, Lord Peter, working with a dentist friend, unravels a well-orchestrated and concealed murder by examining the dental work of the dead man and comparing it to his alleged dental chart. The chart shows an entry for a fused porcelain filling in 1923, but examination shows it to be a cast filling, a kind that did not come into use in England until 1928. Further, the other dental work proved to be makeshift and hurried. The dental evidence was crucial, as all other evidence had been destroyed by fire. The absconding murderer, the dentist who had performed the work, is brought to justice.

P. D. James and Adam Dalgliesh

Dorothy L. Sayers's excellent literary style and sound employment of science and scientific method in her forensic detective fiction rank her high among the writers in this genre. The same is true of P. D. James, whose character Adam Dalgliesh is the investigator in her fine detective mysteries.

James has written 11 novels in this series; several describe the use of forensic methods. Her background as a senior civil servant in the Home Office and the police and criminal departments in London are reflected in her fiction. *Death of an Expert Witness*[14] has several illustrations of how forensic methods are used in criminal investigations. These include deductions based on a doctor's examination of a victim prior to autopsy, forensic biology, and tire mark impressions. The discussion on the function of the document examiner is of particular interest. Descriptions of the work of his laboratory, image enhancement by use of soft X-rays, the dating of papers, and detection of forgery and handwriting analysis make fascinating reading. James's *Devices and Desires*[15] describes how a footprint is found, cast, and then used in a murder investigation.

State-of-the-Art Detective Fiction

Patricia D. Cornwell and Kay Scarpetta

Patricia D. Cornwell is perhaps today's most popular detective mystery writer using forensic methods in her fiction. Like P. D. James's, her stories have a contemporary ring and have moved a long way from the confinement of the country estate settings of Agatha Christie. Cornwell's characters are well-developed and real. Her protagonist in several novels[16] is Kay Scarpetta, M.D., the chief medical examiner of Richmond, Virginia. Scarpetta is quite a lady: a first-class forensic scientist, physician, attorney, and a good pistol shot who also has a warm, loving personality. Her co-worker, police detective Pete Marino, is equally impressive. Cornwell's background includes work as an outstanding crime reporter for the *Charlotte Observer* and more than six years as a computer analyst in the chief medical examiner's office in Richmond.

Cornwell has Kay Scarpetta use the very latest forensic techniques in her work and does not hesitate to give, on occasion, illuminating explanations of forensic methods. With all that, Cornwell maintains an exciting, edge-of-the-chair tale. *Body of Evidence*[16b] demonstrates the use of many new forensic techniques. The character Benton Wesley, who appears in several of Cornwell's novels, shows how he uses psychological profiling to help identify a criminal. Also, there is an excellent discussion on fiber analysis, including the use of a stereoscopic microscope, methods of cleaning individual fibers, and the relationship between fiber and fiber color depending on the fiber's source: garments of various types, rugs, and others.

Cornwell presents a detailed technical discussion of new techniques used to gather evidence at the crime scene in *Post Mortem.*[16a] The use of a laser in an autopsy to obtain trace evidence when examining different tissues and fingerprints on skin helps solve a murder. Cornwell explains how a computer network is used to rapidly identify fingerprints, and there is an excellent discussion of various techniques used in blood grouping, protein analysis, enzyme analysis, and DNA fingerprinting. The use of the scanning electron microscope for identification is another novel technique mentioned.

Cornwell describes computerized networks of criminal identification in *Cruel and Unusual,*[16d] including centralized criminal record exchange, criminal history records, and centralized fingerprint records. Old bloodstains are identified to give a record of what happened during a crime by use of luminol (3-aminophthalhydrazide) to trace the path of bloodstains in the house where the murder occurred. In *All that Remains*[16c] many of the forensic methods mentioned previously are used, plus computer use in drug identification, trace analysis of drugs, and an example of forensic anthropology. Patricia Cornwell is a modern Doyle, not only because of the extraordinary use of forensic science in her detective fiction but also because of the modern tenor of her characters' personalities and social milieu.

Other Modern Detectives

Dilys Winn's *Murder Ink*[17] has two interesting sections. Chapter 11, "The Lab", mentions many cases of doctors who are villains and gives examples of the use of forensic science in criminal investigations. Winn also includes interesting sections on forensic dentistry and anthropology.

In the *Encyclopedia of Mystery and Detection,*[18] edited by Steinbrunner and Penzler, is a section entitled "Scientific Detectives". It mentions many science fiction approaches to murder weapons; Hugo Gernsbach, in his periodicals *Radio News* and *Amazing Stories,* is a major contributor. The section discusses many fictional detectives who use forensic science, largely in short stories in periodicals.

Ian D. Rae, of the chemistry department of Monash University in Victoria, Australia, is the author of a fine review of fictional forensic science, "Dustcoats in Dustjackets".[19] He considers not only forensic chemistry in fiction but also chemists and chemistry in fiction. He mentions J. J. Connington, the author of several fictional books that incorporate forensic chemistry. Connington was the pen name of A. W. Stewart, professor of chemistry at Belfast University.

In this chapter, I have covered some of the highlights of forensic science in detective fiction, from imaginative predictions of things to come to the common techniques of the crime laboratory. Recently, a new dimension has been added to the crime novel: authors are appealing especially to the scientifically trained reader. These authors show that behind the legal scene is a body of hard science that not only establishes guilt but defends the innocent. No doubt other worthy authors have been omitted; I would be delighted to hear from readers who know of other writers or other books that deserve mention in future editions of this book.

References

1. Benton, W. *Encyclopedia Britannica* **1965,** *2,* 294.
2. There are many fine complete collections of Sherlock Holmes fiction. My starting point was Doyle, A. C. *The Complete Sherlock Holmes*, Vols. I and II; Doubleday: Garden City, NY, 1939. A delightful (and completely illustrated, mostly by Sidney Paget) and well-annotated edition is *The Annotated Sherlock Holmes*, Vols. I and II; Ed. William S. Baring-Gould; Clarkson N. Potter: New York, 1967. The most recent collection is *The Oxford Sherlock Holmes*, 9 vols; General Ed. O. D. Edwards; Oxford University Press: Oxford, 1993. This edition has excellent introductions to the stories and relevant bibliographies.
3. Liebow, E. M. *Dr. Joe Bell, Model for Sherlock Holmes;* Bowling Green University Popular: Bowling Green, OH, 1982.
4. Lee, H. C. In *Forensic Science Handbook;* Safirstein, R., Ed. Prentice-Hall: Englewood Cliffs, NJ, 1982; Vol. 1, pp 272, 284.
5. Eco, U. *The Name of the Rose;* Harcourt-Brace: New York, 1983.
6. Haft, A. J.; White, J. G.; White, R. J. *The Key to the Name of the Rose;* Ampersand Associates: Harrington Park, NJ, 1987; p 27.
7. Freeman, R. A. *The Best Dr. Thorndyke Detective Stories;* Dover: New York, 1973.
8. *Encyclopedia of Mystery and Detection;* Steinbrunner, C.; Penzler, O., Eds.; McGraw-Hill: New York, 1976; p 387.
9. Sayers, D. L.; Eustace, R. *The Documents in the Case;* Avon: New York, 1968.
10. Foster, N. "Strong Poison: Chemistry in the Works of Dorothy L. Sayers." In *Chemistry and Crime: From Sherlock Holmes to Today's Courtroom;* Gerber, S. M., Ed.; American Chemical Society: Washington, D. C., 1983, p 18ff.
11. Sayers, D. L. *Strong Poison;* Harper and Row: New York, 1987.
12. Sayers, D. L. *Unnatural Death* (originally published as *The Dawson Pedigree*); Harper and Brothers: New York, 1955.
13. Sayers, D. L. *In the Teeth of the Evidence and Other Mysteries;* Harper and Row: New York, 1987 (originally published by Harcourt, Brace, Jovanavich, 1940).
14. James, P. D. *Death of An Expert Witness;* Charles Scribners Sons: New York, 1977.
15. James, P. D. *Devices and Desires;* Alfred A. Knopf: New York, 1990.
16. Cornwell, P. D. (a) *Post Mortem;* 1990. (b) *Body of Evidence;* 1991. (c) *All That Remains;* 1992. (d) *Cruel and Unusual;* 1993. (e) *The Body Farm;* 1994. (f) *Original Sin;* 1995. Charles Scribners Sons: New York.

17. *Murder Ink: The Mystery Reader's Companion;* Winn, D., Ed.; Workman: New York, 1977.
18. Ref. 8 above, p 356, 7.
19. Rae, I. D. "Dustcoats in Dustjacket"; *Chemistry in Britain*, 1983; July, p 565–568.

Index

Jacket design: Rhonda Rawlings
Text design and typesetting by Betsy Kulamer, Washington, DC
Printing and binding by Maple Press Company, York, PA